Algebra

Algebra Parts I, II, and III combined

Lesson/Practice Workbook
for Self-Study and Test Preparation

Build Your Self-Confidence and Enjoyment of Math!
Comprehensive Solutions Manual Sold Separately

Aejeong Kang

MathRadar

Send all inquiries to:
MathRadar, LLC
5705 Spring Hill Dr.
Mckinney, Texas 75072

Visit www.mathradar.com for more information and a sneak preview of the MathRadar series of math books.

Send inquires via email: info@mathradar.com

Algebra: Lesson/Practice Workbook for Self-Study and Test Preparation
(Algebra Parts I, II, and III combined)

ISBN-13: 978-0-9960450-1-8

ISBN-10: 0996045015

Printed in the United States of America.

Preface

I wrote these books because I am a mother and I have a strong academic background in mathematics. I have a BS degree in Mathematics and Master's degree in Mathematics as well. I have completed Ph.D. program in Biostatistics.

After receiving the big blessing of our first child, a daughter, I decided to forgo my personal career goals to become a full-time mother. When our daughter entered 7th grade, that meant lots of help with her study of math-my passion. However, I struggled to find good math books that would help her understand difficult concepts both clearly and quickly. About two years ago, I talked with my husband and my kids (now I have 2 children 8th grader, Nichole and 1st grader, Richard) about an idea that it would be better to write math books myself at least for my kids because I really want my kids study math with best books. After the conversation, I decided that the best way to help my children was by writing math books for them myself. They wholeheartedly agreed.

That's why I've been able to pour all my knowledge, energy, and soul into Mathradar Series. Because I'm a mom, I would do anything for my children. Thanks to my family's endless support, I wrote them, designed for use in junior high, high-school, and advanced high-school mathematics.

And that would have been the end of my journey, but my husband and children insisted that I share my work outside of our family. They encouraged me to make my work available to other parents looking, as I was, for well-written, great mathematics books for their children.

So I finally decided to publish these books. I do so with the hope that they will help your children find success and confidence in learning and studying mathematics.

But I would never have begun or finished this project without the support of my family. Kyungwan, Nichole, and Richard, you are my world. Thank you.

Aejeong Kang

Introduction

After reading several pages of explanation/description about a certain mathematical concept, you still don't get it.

You have worked on many related problems to understand mathematical concepts, but you still feel completely lost in the mathematical jungle.

You bought a math book with good reviews, but it only offers short answers without detailed solutions. You feel confused and frustrated.

You've tried multiple learning math books, but you've still not getting good grades in math. It seems like math is just not for you.

If any one of these situation sound familiar, the MathRadar series will help you escape!

Everyone has different learning abilities and academic skill. MathRadar series is written and organized with emphasis on helping each individual study mathematics at his/her own pace. Each book consists of clean and concise summaries, callouts, additional supporting explanations, quick reminders and/or shortcuts to facilitate better understanding. Each concept is thoroughly explained with step-by-step instruction and detailed proofs.
With the numerous examples and exercises, students can check their comprehension levels with both basic and more advanced problems.

Carry the MathRadar series with you!
Work on them anytime and anywhere!
Finally, you can start to enjoy mathematics!

Whether you are struggling or advanced in your math skills, the MathRadar series books will build your self-confidence and enjoyment of math.

I hope Math Radar is what you need and will be a great tool for your hard work.
Your comments or suggestions are greatly appreciated.
Please visit my website at www. mathradar.com or email me at ae-jeong@mathradar.com
Thank you very much. And remember, math can be fun!

TABLE OF CONTENTS

Chapter 3. Equations

Chapter 4. Inequalities

Chapter 5. Functions

Chapter 6. Fractions and Other Algebraic Expressions

Chapter 7. Monomials and Polynomials

Chapter 8. Systems of Equations

Chapter 9. Systems of Inequalities

Chapter 10. Linear Functions

Chapter 11. The Real Number System

Chapter 12. Factorization

Chapter 13. Quadratic Equations

Chapter 14. Rational Expressions (Algebraic Functions)

Chapter 15. Quadratic Functions

Chapter 16. Basic Statistical Graphs

Chapter 17. Descriptive Statistics

Chapter 18. The Concept of Sets

Chapter 19. Probability

Answer Key

Index

Chapter 1. The Natural Numbers

1-1 Factors and Multiples

1. Definition

Given $a = bc$ for any integers $a, b \neq 0,\ c$,

a is called a *multiple* of b and b is called a *factor* of a.

---

① Natural numbers are the numbers we count with .

(starting with 1 and going up by ones) : $1, 2, 3, 4, 5, \cdots$

② Integers

$$\begin{cases} 1,\ 2,\ 3,\ \cdots\ ; \text{positive (: natural numbers)} \\ \quad 0 \quad ; \text{zero} \\ -1, -2, -3, \cdots ; \text{negative} \end{cases}$$

③ Whole numbers $\begin{cases} 0 \\ \text{positive} \end{cases}$

└─ All positive numbers and a zero.

For any natural numbers $a,\ b,$ and c,

$a = bc \ \Rightarrow\ a$ is a multiple of b or multiple of c .

b and c are factors of a.

1 is the smallest factor of all natural

numbers and 1 has only one factor , 1.

← See Chapter 2 for more information!

Note :

(1) $0 = 1 \times 0,\ 0 = 2 \times 0,\ 0 = 3 \times 0,\ \cdots \cdots\ ;\ 0$ *is a multiple of* $1,\ 2,\ 3, \cdots\cdots$ *(natural numbers)*

and $1,\ 2,\ 3,\ \cdots$ *(natural numbers) are factors of* 0.

But 0 *is not a multiple of* 0 *or* 0 *is not a factor of* 0 *because factor and multiple are not defined when*

$b = 0$.

Therefore, 0 *is not a factor of any numbers but* 0 *is multiple of any natural numbers.*

(2) $\dfrac{3}{4} = \dfrac{1}{4} \times 3,$ *but* $\dfrac{3}{4}$ *is not a multiple of* $\dfrac{1}{4}$ *or* $\dfrac{1}{4}$ *is not a factor of* $\dfrac{3}{4}$

because factors and multiples are defined by integers.

(3) $6 = -2 \times -3,$ *but negative numbers are excluded, conventionally.*

Thus, factors and multiples are applied on whole numbers only.

(4) *If* $\dfrac{a}{n},\ \dfrac{b}{n}$ *are natural numbers ,*

$$\frac{nc}{n} = c ,\ \frac{nd}{n} = d$$
$$\frac{ac}{a} = c ,\ \frac{bd}{b} = d$$

then $a = n \times c,\ b = n \times d$ *for reduction. So,* n *is the common factor of a and b.*

If $\dfrac{n}{a},\ \dfrac{n}{b}$ *are natural numbers,*

then $n = a \times c,\ n = b \times d$ *for reduction. So,* n *is the common multiple of a and b.*

2. Divisibility Tests

A whole number is divisible (0 remainder) by each of its factors.

To determine whether large numbers are divisible by smaller numbers, divisibility tests are the fastest methods. When a number is :

(1) Divisible by 2

Its last digit is 0 or an even number.

(2) Divisible by 3

The sum of all of the digits is a multiple of 3.

(3) Divisible by 4

Its last two digits are 00 or a multiple of 4.

(4) Divisible by 5

Its last digit is either 0 or 5.

(5) Divisible by 6

It is divisible by both 2 and 3.

(6) Divisible by 7

Separate the last digit from the whole number and double the last digit. Then subtract the doubled digit from the rest of the number. If the difference is not obvious, repeat this method with the difference. See an example below.

(7) Divisible by 8

Its last three digits are 000 or a multiple of 8.

(8) Divisible by 9

The sum of all of the digits is a multiple of 9.

(9) Divisible by 10

Its last digit is 0.

(10) Divisible by 11

The difference of the sum of odd digits and the sum of even digits is 0 or a multiple of 11.

Is 4564 divisible by 7 ?

(∵ separate the last digit ⇒ 456 and 4

double the last digit ⇒ $4 \times 2 = 8$

subtract the doubled digit from 456 ⇒ $456 - 8 = 448$

Now, let's repeat this.

separate :⇒ 44 and 8

double :⇒ $8 \times 2 = 16$

subtract :⇒ $44 - 16 = 28$

Since 28 is divisible by 7, 4564 is divisible by 7.)

Is 31542 divisible by 11 ?

:Odd digits

$(3 + 5 + 2) - (1 + 4) = 10 - 5 = 5$

∴ Not divisible by 11

Is 42372 divisible by 11 ?

$(4 + 3 + 2) - (2 + 7) = 9 - 9 = 0$

∴ Divisible by 11

3. Prime Factorization

(1) Prime Number

A natural number is called *prime* if it has no natural numbers as factors except itself and 1.

Example : 2, 3, 5, 7, 11, 13, 17, 19, $\cdots\cdots$

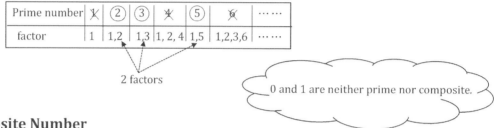

2 is the only even prime number.

Note : ① *2 is the smallest prime number.*

② *All of the prime numbers have only 2 factors : 1 and itself.*

③ *1 is not considered as a prime number because it has only 1 factor: itself.*

Prime number	~~1~~	②	③	~~4~~	⑤	~~6~~	$\cdots\cdots$
factor	1	1,2	1,3	1,2,4	1,5	1,2,3,6	$\cdots\cdots$

2 factors

0 and 1 are neither prime nor composite.

(2) Composite Number

A natural number is called *composite* if it is greater than 1 and not a prime.

Example: 4, 6, 8, 9, 10, 12, 14, 15, 16, 18, 20, $\cdots\cdots$

Note : All composite numbers have more than 2 factors.

(3) Prime Factorization

A natural number can be expressed as a product of prime numbers in exactly one way, not considering the order of the factors. Expressing a composite number as a product of primes is called *prime factorization*.

1) Factor Tree (Tree Diagram) :

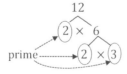

$12 = 2 \times 2 \times 3 = 2^2 \times 3$: product of prime numbers

2) Shortcut Division :

$$2 \,\overline{)\, 12}$$
prime $\longleftarrow$ $2 \,\overline{)\, 6}$
3

12= 1 ×12=2× 6 = 3 × 4
Factors: 1, 2, 3, 4, 6, 12
Prime factors : 2, 3
Prime factorization : $2^2 \times 3$

$12 = 2 \times 2 \times 3 = 2^2 \times 3$: Product of prime numbers

Note : Divide the number by a prime number until the last row is a prime number.

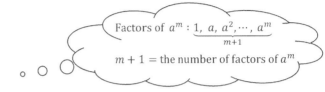

Factors of a^m : $\underbrace{1, \ a, \ a^2, \cdots, \ a^m}_{m+1}$

$m + 1 =$ the number of factors of a^m

4. The Number of Factors

For any natural numbers m and n, if a natural number N is factored by $a^m \times b^n$, where a and b are different primes, then the number of factors of N is $(m + 1) \times (n + 1)$.

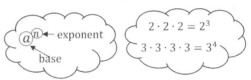

Factors of 12 are
1, 2, 3, 4, 6, 12 .

Example $\quad 12 = 2^2 \times 3 = 2^2 \times 3^1$

$\Rightarrow$ The number of factors of 12 is $\underbrace{(2 + 1)}_{} \times \underbrace{(1 + 1)}_{} = 3 \times 2 = 6.$

$\quad\quad$ number of $\quad\quad$ number of
$\quad\quad$ factors of 2^2 $\quad$ factors of 3

5. Exponents

For any number $a(\neq 0)$, an *exponential form* for a multiplication of n times repeated factors of a is defined by a^n, where a is called a *base* and n is called an *exponent* (or *power*).

All numbers can be written as exponential forms.

For example, $1 = 1^1$, $2 = 2^1$, $\cdots\cdots$

a^{n} $\leftarrow$ exponent
base

$2 \cdot 2 \cdot 2 = 2^3$
$3 \cdot 3 \cdot 3 \cdot 3 = 3^4$

1-2   Greatest Common Factor and Least Common Multiple

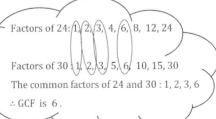

Factors of 24: 1, 2, 3, 4, 6, 8, 12, 24
Factors of 30 : 1, 2, 3, 5, 6, 10, 15, 30
The common factors of 24 and 30 : 1, 2, 3, 6
$\therefore$ GCF is 6 .

1. Greatest Common Factor (GCF)

(1) Definition

GCF is the *greatest common factor* of two or more given numbers.

GCF is the largest factor that divides into all given numbers evenly.

(2) Finding Method

1) Using prime factorization

Step 1. Generate the prime factorization of each number.

Step 2. Find the lowest power of each factor.

Step 3. Multiply the common factors that have the lowest power.

Example Find the GCF of 24 and 30.

$24 = 2^3 \times 3$

$30 = 2 \times 3 \times 5$

The common prime factors of lowest power are 2^1 and 3^1.

$\therefore$ GCF is $2^1 \times 3^1 = 6$.

2) Using common factors

Step 1. Divide each number by a common factor which is not 1.

Step 2. Keep dividing until there is no common factor except 1.

Step 3. Multiply all the common factors which divided each number (multiply all the divisors).

Example

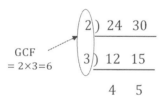

$$\begin{array}{r|cc} 2 & 24 & 30 \\ \hline 3 & 12 & 15 \\ \hline & 4 & 5 \end{array}$$

GCF
$= 2 \times 3 = 6$

> Start with the smallest common factor and move to larger numbers when dividing the given numbers.

Step 1.   Divide 24 and 30 by a common factor 2.

Step 2.   The quotients 12 and 15 have common factor 3.

So, keep dividing 12 and 15 by 3. The quotients 4 and 5 have only 1 as a common factor. Then, stop dividing here.

Step 3.   GCF= $2 \times 3 = 6$

2. Least Common Multiple (LCM)

(1) Definition

LCM is the *least common multiple* of two or more given numbers.

LCM is the smallest non-zero factor that is a multiple of all given numbers.

> Multiple of 12 : 0, 12, 24, 36, 48, 60, 72, ······
>
> Multiple of 18 : 0, 18, 36, 54, 72, ······
>
> Common multiples of 12 and 18 : 0, 36, 72, ······
>
> $\therefore$ LCM is 36 (; the smallest non-zero factor).

(2) Finding Method

1) Using prime factorization

Step 1. Generate the prime factorization of each number.

Step 2. Find the highest power of each factor.

Step 3. Multiply all the factors that have the highest power.

Example   Find the LCM of 12, 18, and 30.

$12 = 2^2 \times 3$

$18 = 2 \times 3^2$

$30 = 2 \times 3 \times 5$

Factors that have the highest power of each factor are $2^2, 3^2$ and 5.

$\therefore \text{LCM} = 2^2 \times 3^2 \times 5 = 180$

2) Using common factors

Step 1. Divide each number by a common factor which is not 1.

Step 2. Keep dividing until there is no common factor except 1.

If divide three numbers and there is no common factor of all three numbers,

then divide them by the common factor of two numbers

and leave the other remaining number as it is.

Step 3. Multiply all the divisors and the numbers on the last row.

Example 1

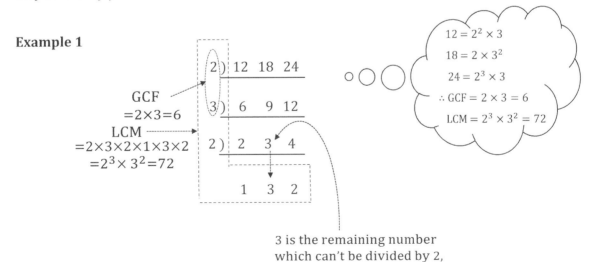

3 is the remaining number
which can't be divided by 2,
the divisor of the other two numbers, 2 and 4.

Step 1. Divide all three numbers (12, 18, and 24) by their common factor 2.

Step 2. Divide the resulting three numbers (6, 9, and 12) by their common factor 3.

Divide only the two numbers 2 and 4 by the common factor 2, and leave the remaining number 3 as it is.

Step 3. Multiply all the divisors (2, 3, and 2) and the numbers on the last row (1, 3, and 2)

$$\therefore LCM = 2 \times 3 \times 2 \times 1 \times 3 \times 2 = 2^3 \times 3^2 = 72$$

Example 2

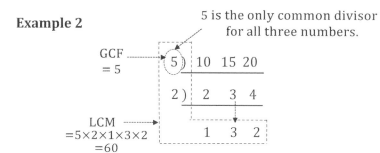

Example 3

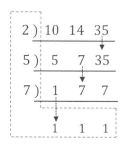

GCF =1
($\because$ there is no common divisor
except 1 for all three numbers.)

LCM
$= 2 \times 5 \times 7 \times 1 \times 1 \times 1$
$= 70$

3. The Relationship of GCF and LCM

Let A and B be two natural numbers.

If GCF of A and B is G and LCM of A and B is L, then

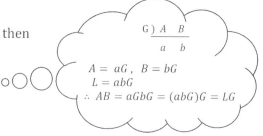

(1) $A = aG$, $B = bG$

(2) $L = abG$

(3) $AB = LG$

for any two numbers a and b which don't have any common factor except 1.

Example 1 $A = 12,\ B = 18$

$$\Rightarrow \quad 2\ )\ \underline{12\quad 18}$$

$$3\ )\ \underline{6\quad 9}$$

$$2\quad 3$$

$\therefore$ GCF$= 2 \times 3 = 6$

LCM$= 2 \times 3 \times 2 \times 3 = 2^2 \times 3^2$

① $12 = 2 \times 6\ (A = aG)$, $18 = 3 \times 6\ (B = bG)$

② $2^2 \times 3^2 = 2 \times 3 \times 6\ (L = abG)$

③ $12 \times 18 = 2^2 \times 3^2 \times 6\ (AB = LG)$

Example 2 $A = 36xy^2,\ B = 60x^2y$

$\Rightarrow A = 2^2 \cdot 3^2 \cdot x \cdot y^2$ and $B = 2^2 \cdot 3 \cdot 5 \cdot x^2 \cdot y$

$\therefore$ GCF of A and $B = 2^2 \cdot 3 \cdot x \cdot y = 12xy$

LCM of A and $B = 2^2 \cdot 3^2 \cdot 5 \cdot x^2 \cdot y^2 = 180x^2y^2$

① $36xy^2 = 3y \cdot 12xy\ (A = aG)$, $60x^2y = 5x \cdot 12xy\ (B = bG)$

② $180x^2y^2 = 3y \cdot 5x \cdot 12xy\ (L = abG)$

③ $36xy^2 \cdot 60x^2y = 180x^2y^2 \cdot 12xy = 2160x^3y^3\ (AB = LG)$

Exercises

#1. Find all the factors and multiples of each number.

(1) 4   (5) 1

(2) 7   (6) 15

(3) 12   (7) 18

(4) 36   (8) 0

#2 In each of the following, determine whether the statement is true or false.

(1) 1 is a prime number.

(2) A prime number is an odd number.

(3) A prime number has 2 factors.

(4) All the natural numbers are considered as prime numbers and composite numbers.

(5) 0 is a factor of 3.

(6) The product of two prime numbers is a composite number.

(7) A composite number is an even number.

(8) Each natural number has itself as a factor.

(9) The smallest composite number is 4.

(10) All multiples of 3 are composite numbers.

(11) The prime number of 9 is 3^2.

(12) The prime factorization of 36 is $2^2 \times 9$.

#3 Find the value of $a + b$ for any two natural numbers a, b which satisfy the following

(1) $3^a = 729$, $4^b = 64$ (3) $360 = 2^3 \cdot 3^a \cdot 5^b$

(2) $2^a \cdot 3^b = 324$ (4) $32 = 2^a$, $108 = 2^2 \cdot 3^b$

#4 Find the prime factors for the following

(1) 24   (4) 60

(2) 63   (5) 210

(3) 56   (6) 100

#5 For any natural numbers a and b, where a is the smallest number possible,
find the value of $a + b$.

(1) $8a = b^2$

(2) $48a = b^2$

(3) $56a = b^2$

(4) $360a = b^2$

(5) $\frac{32}{a} = b^2$

(6) $\frac{120}{a} = b^2$

(7) $\frac{150}{a} = b^2$

(8) $\frac{135}{a} = b^2$

#6 Find the number of factors for the following

(1) 15

(2) 24

(3) 36

(4) 96

(5) 225

#7 The number of factors for the following two numbers are the same. Find the value of $a + b$ for
the natural numbers a and b.

(1) 180 and $30 \cdot 3^a \cdot 5^b$

(2) 72 and $12 \cdot 2^a \cdot 3^b$

(3) 216 and $4 \cdot 2^{a-2} \cdot 3^b \cdot 5$

#8 Find the GCF and LCM for the following

(1) $2^2 \cdot 3 \cdot 5$ and $2^3 \cdot 3^2 \cdot 5 \cdot 7$

(2) $2^4 \cdot 3 \cdot 5^2 \cdot 11$ and $2^2 \cdot 3^3 \cdot 7$

(3) $2^3 \cdot 3^2 \cdot 5$, $2 \cdot 3^3 \cdot 5^2 \cdot 7$, and $2^2 \cdot 3^4 \cdot 7$

(4) 90, $3^3 \cdot 5 \cdot 7$, and $2^2 \cdot 3 \cdot 7^2$

#9 Find the value of $a + b$ for the following

(1) The GCF and LCM for $2^a \cdot 3^2$ and $2^3 \cdot 3^b \cdot 5^2$ are $2^3 \cdot 3$ and $2^5 \cdot 3^2 \cdot 5^2$, respectively.

(2) The GCF and LCM for $2^3 \cdot 3^a \cdot 5$ and $2 \cdot 3^4 \cdot 5^2 \cdot 7$ are $2 \cdot 3^3 \cdot 5$ and $2^b \cdot 3^4 \cdot 5^2 \cdot 7$, respectively.

#10   Find all the values of n which would make the following fractions natural numbers

(1) $\dfrac{24}{n}$, $\dfrac{36}{n}$

(2) $\dfrac{12}{n}$, $\dfrac{30}{n}$

(3) $\dfrac{n}{6}$, $\dfrac{n}{8}$

(4) $\dfrac{n}{9}$, $\dfrac{n}{15}$

(5) $\dfrac{35}{6}n$, $\dfrac{20}{9}n$

#11   Solve the following

(1) The GCF and LCM for two natural numbers N and 48 are 12 and 144, respectively. Find the number N.

(2) The product of two natural numbers A and B is 480 and their GCF is 4. Find their LCM.

(3) The GCF and LCM for two natural numbers A and B are 6 and 210, respectively. Find the value of AB.

(4) The GCF and LCM for two natural numbers A and B, where A is a multiple of B, are 8 and 48, respectively. Find the value of $A + B$.

Chapter 2. Integers and Rational Numbers

2-1 Integers

1. Classification of Integers

The *integers* consist of the positive integers (natural numbers), zero, and the negative integers (negatives of the natural numbers).

(1) Positive Integers (Natural numbers): The numbers we count with.

Integers which are greater than 0.

$+1, +2, +3, +4, \cdots\cdots$ or $1, 2, 3, 4, \cdots\cdots$

(2) Negative Integers: The natural numbers with negative signs in front.

Integers which are less than 0.

$-1, -2, -3, -4 \cdots\cdots$

(3) Zero: Integer which is neither positive nor negative.

Non-positive or non-negative.

$$\text{Integers} \begin{cases} \text{Positive integers} \\ \text{Negative integers} \\ \text{Zero} \end{cases}$$

$$\text{Whole numbers} = \begin{cases} \text{Positive integers} \\ \text{Zero} \end{cases}$$

Note : ① *Number line*

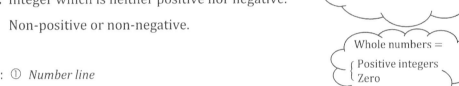

Integers

$$\cdots \ -4 \ -3 \ -2 \ -1 \ \ 0 \ \ 1 \ \ 2 \ \ 3 \ \ 4 \ \cdots$$

Number line

Negative integers ⟵   (Origin)   ⟶ Positive integers

② *Whole numbers are non-negative integers.*

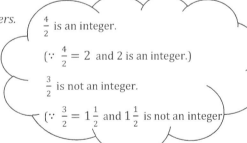

$\frac{4}{2}$ is an integer.

$(\because \frac{4}{2} = 2$ and 2 is an integer.$)$

$\frac{3}{2}$ is not an integer.

$(\because \frac{3}{2} = 1\frac{1}{2}$ and $1\frac{1}{2}$ is not an integer$)$

2. Magnitude of Integers

(1) Positive integers are greater than 0.

(2) Negative integers are less than 0.

(3) Negative integers $< 0 <$ Positive integers.

That is, positive integers are greater than negative integers.

(4) When comparing two positive integers, the positive integer farthest away from 0 is larger, and when comparing two negative integers, the negative integer closest to 0 is larger.

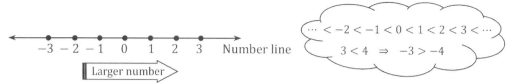

The number on the right side is larger than the number on the left side on a number line.

3. Operations of Integers

(1) Adding Integers

For any two integers a and b,

1) If the signs of a and b are the same:

> ① Pretend both numbers are positive.
>
> ② Add two numbers, ignoring their signs .
>
> ③ Place their sign in front of the sum.

If a and b are both positive, then the sum is positive.

$a > 0,\ b > 0\ \Rightarrow\ a + b > 0$

Example $\quad 2 + 4 = +(2+4) = +6 = 6 > 0$

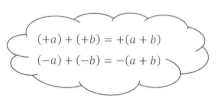

If a and b are both negative, then the sum is negative.

$a < 0,\ b < 0\ \Rightarrow\ a + b < 0$

Example $\quad (-2) + (-4) = -(2+4) = -6 < 0$

2) If the signs of a and b are different:

> ① Pretend both numbers are positive.
>
> ② Subtract the smaller number from the larger one, ignoring their signs.
>
> ③ Place the sign from the larger number.

Example $\quad 2 + (-4) = -(4-2) = -2$

Add: -5 and -6 (same sign)

$\Rightarrow 5 + 6$ (just add)

$\Rightarrow -11$ (use their sign)

Add: -5 and 6 (different signs)

$\Rightarrow 6 - 5$ (just subtract)

$\Rightarrow +1$ (use the sign of 6 ($\because 6 > 5$))

3) If one of the numbers is zero:

Any number added to zero is that number.

$a + 0 = a,\ 0 + b = b,\ (-a) + 0 = -a,\ 0 + (-b) = -b$

Example $\quad 2 + 0 = 2,\ 0 + (-2) = -2$

4) Adding any number to its opposite equals zero.

$a + (-a) = 0, \ (-a) + a = 0$

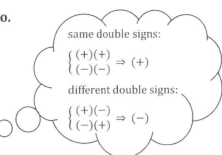

same double signs:

$\begin{cases} (+)(+) \\ (-)(-) \end{cases} \Rightarrow (+)$

different double signs:

$\begin{cases} (+)(-) \\ (-)(+) \end{cases} \Rightarrow (-)$

Example $2 + (-2) = 0, \ (-2) + 2 = 0$

(2) Subtracting Integers

Convert the double signs (two signs next to one another) into a single sign.

1) | **If the double signs are the same $\Rightarrow +$ (positive sign)**

$a + (+b) = a + b, \quad a - (-b) = a + b$

Example $2 - (-4) = 2 + 4 = 6$

$(-)(-) = (+)$

Adding two numbers
with different signs

$-2 - (-4) = -2 + 4 = +(4 - 2) = +2 = 2$

2) | **If the double signs are different $\Rightarrow -$ (negative sign)**

$a + (-b) = a - b, \quad a - (+b) = a - b$

Example $2 - (+4) = 2 - 4 = -(4 - 2) = -2$

$(-)(+) = (-)$

Aadding two numbers
with different signs

$-2 - (+4) = -2 - 4 = -(2 + 4) = -6$

Adding two numbers
with the same signs

3) | **If one of the numbers is zero:**

Zero does not have any effect.

If there are double signs, then convert them into a single sign.

$a - 0 = a, \ -a - 0 = -a, \ 0 - (+a) = 0 - a = -a, \ 0 - (-a) = 0 + a = a$

Example

$2 - 0 = 2, \qquad -2 - 0 = -2, \qquad 0 - (+2) = 0 - 2 = -2, \qquad 0 - (-2) = 0 + 2 = 2$

(3) Multiplying Integers

For any two integers a and b,

1) If the signs of a and b are the same, whether they are both positive or both negative:

> ① Pretend both numbers are positive.
>
> ② Multiply the two numbers, ignoring their signs.
>
> ③ Place the positive sign in front of the product.

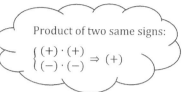

Product of two same signs:
$$\begin{cases} (+) \cdot (+) \\ (-) \cdot (-) \end{cases} \Rightarrow (+)$$

If a and b are both positive, then the product will be positive as well.

$$a > 0, \ b > 0 \ \Rightarrow \ a \times b > 0$$

If a and b are both negative, then the product will be positive.

$$a < 0, \ b < 0 \ \Rightarrow \ a \times b > 0$$

Example $2 \times 4 = +(2 \times 4) = +8 = 8 > 0$

$\qquad\quad (-2) \times (-4) = +(2 \times 4) = +8 = 8 > 0$

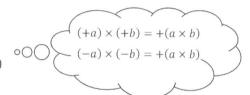

$(+a) \times (+b) = +(a \times b)$
$(-a) \times (-b) = +(a \times b)$

2) If the signs of a and b are different :

> ① Pretend both numbers are positive.
>
> ② Multiply the two numbers, ignoring their signs.
>
> ③ Place the negative sign in front of the product.

Product of two different signs:
$$\begin{cases} (+) \cdot (-) \\ (-) \cdot (+) \end{cases} \Rightarrow (-)$$

$$a > 0, \ b < 0 \ \Rightarrow \ a \times b < 0$$

$$a < 0, \ b > 0 \ \Rightarrow \ a \times b < 0$$

Example $2 \times (-4) = -(2 \times 4) = -8 < 0$

$\qquad\quad (-2) \times 4 = -(2 \times 4) = -8 < 0$

The number of $(-)$ sign is even

$\Rightarrow (+)$: positive sign

EX. $(-1)(-2)(+3) = +6$

$\qquad (-1)(-2)(-3)(-4) = +24$

The number of $(-)$ sign is odd

$\Rightarrow (-)$: negative sign

EX. $(+1)(-2)(+3) = -6$

$\qquad (-1)(-2)(-3) = -6$

3) If one of the numbers is zero :

> Any number multiplied by zero equals zero.
>
> $a \times 0 = 0, \ \ 0 \times b = 0$

Example

$2 \times 0 = 0, \ \ 0 \times (-2) = 0$

(4) Dividing Integers

For any two integers a and b,

1) If the signs of a and b are the same, whether they are both positive or both negative:

> ① Pretend both numbers are positive.
>
> ② Divide the two numbers, ignoring their signs.
>
> ③ Place the positive sign in front of the division.

If a and b are both positive, then the division will be positive as well.

$$a > 0,\ b > 0 \ \Rightarrow\ \frac{b}{a} > 0,\ \frac{a}{b} > 0$$

If a and b are both negative, then the division will be positive as well.

$$a < 0,\ b < 0 \ \Rightarrow\ \frac{b}{a} > 0,\ \frac{a}{b} > 0$$

Example $\dfrac{2}{4} = \dfrac{1}{2} > 0, \quad \dfrac{4}{2} = 2 > 0$

$\qquad\qquad \dfrac{-2}{-4} = \dfrac{1}{2} > 0, \quad \dfrac{-4}{-2} = 2 > 0$

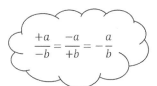

$$\frac{+a}{-b} = \frac{-a}{+b} = -\frac{a}{b}$$

2) If the signs of a and b are different:

> ① Pretend both numbers are positive.
>
> ② Divide the two numbers, ignoring their signs.
>
> ③ Place the negative sign in front of the division.

$$a > 0,\ b < 0 \ \Rightarrow\ \frac{b}{a} < 0,\ \frac{a}{b} < 0$$

$$a < 0,\ b > 0 \ \Rightarrow\ \frac{b}{a} < 0,\ \frac{a}{b} < 0$$

Example $\dfrac{2}{-4} = -\dfrac{1}{2} < 0, \quad \dfrac{-4}{2} = -2 < 0$

$\qquad\qquad \dfrac{-2}{4} = -\dfrac{1}{2} < 0, \quad \dfrac{4}{-2} = -2 < 0$

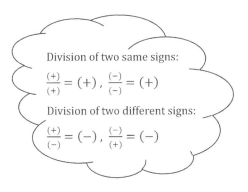

Division of two same signs:

$\dfrac{(+)}{(+)} = (+),\ \dfrac{(-)}{(-)} = (+)$

Division of two different signs:

$\dfrac{(+)}{(-)} = (-),\ \dfrac{(-)}{(+)} = (-)$

3) If one of the numbers is zero:

Zero divided by any number (except 0) is zero.

$$\frac{0}{a} = 0 , \qquad \frac{0}{-a} = 0$$

Division by zero is undefined.

$$\frac{a}{0} = \text{Undefined} , \qquad \frac{-a}{0} = \text{Undefinded}$$

For a number $a(\neq 0)$,

① If $\frac{0}{a} = x \Rightarrow 0 = a \cdot x$

Since $a \neq 0$, $x = 0$. $\therefore \frac{0}{a} = 0$

② If $\frac{a}{0} = x \Rightarrow a = 0 \cdot x$

Since $a \neq 0$ and $0 \cdot x = 0$,

$a = 0 \cdot x$, $a \neq 0$ is not possible.

$\therefore \frac{a}{0}$ is undefined.

Note:

① $\frac{0}{0} = a \Leftrightarrow 0 = a \cdot 0$: *This is always true for any real number a.*

② $\frac{a}{a} = 1$ *for any* $a(\neq 0)$

③ *Zero may never appear in the denominator of a fraction*, *but* $\frac{0}{a} = 0$ *for any* $a(\neq 0)$.

④ $ab = 0$ *for real numbers a and b* $\Rightarrow a = 0$ *or* $b = 0$

($\because$ *case* 1. *when* $a = 0$, $ab = 0 \cdot b = 0$

$ab = 0 \Leftrightarrow$ One of the two numbers is zero.

case 2. *when* $a \neq 0$,

Since $a \neq 0$, $\frac{1}{a}$ *is defined.*

Since $ab = 0$, *multiply* $\frac{1}{a}$ *to both sides of the equal sign in the expression*

(Keep the equation balanced). Then, $\frac{1}{a} \cdot (ab) = \frac{1}{a} \cdot 0$ $\therefore b = 0$)

4. Properties of Integers

(1) Properties of Operations

For any integers a, b, and c,

1) Commutative property (for addition and multiplication)

$$a + b = b + a, \qquad a \cdot b = b \cdot a$$

2) Associative property (for addition and multiplication)

$$(a + b) + c = a + (b + c), \qquad (a \cdot b) \cdot c = a \cdot (b \cdot c)$$

3) Distributive property (for multiplication and division)

$$a \cdot (b + c) = (a \cdot b) + (a \cdot c), \qquad a \cdot (b - c) = (a \cdot b) - (a \cdot c)$$

$$(a + b) \cdot c = (a \cdot c) + (b \cdot c), \qquad (a - b) \cdot c = (a \cdot c) - (b \cdot c)$$

$$\frac{a+b}{c} = \frac{a}{c} + \frac{b}{c} , \ c \neq 0, \qquad \frac{a-b}{c} = \frac{a}{c} - \frac{b}{c} , \ c \neq 0$$

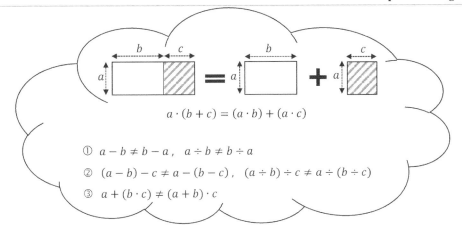

$$a \cdot (b + c) = (a \cdot b) + (a \cdot c)$$

① $a - b \neq b - a$, $a \div b \neq b \div a$

② $(a - b) - c \neq a - (b - c)$, $(a \div b) \div c \neq a \div (b \div c)$

③ $a + (b \cdot c) \neq (a + b) \cdot c$

(2) Order of Operations

Step 1. Remove parentheses ( ) or other enclosure marks (braces { }, brackets [ ],
absolute value signs | |, etc.) in order from the innermost enclosure marks to the
outermost ones.

Step 2. Exponents (Powers)

Step 3. $\times$, $\div$ (Compute in order from left to right, no matter which operation comes first.)

Step 4. $+$, $-$ (Compute in order from left to right, no matter which operation comes first.)

5. Absolute Values

An *absolute value* is the distance in units from a number to the origin (or zero) on a number line
and is expressed as a positive quantity.

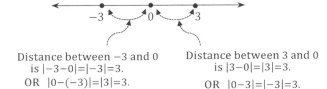

Distance between -3 and 0
is $|-3-0| = |-3| = 3$.
OR $|0-(-3)| = |3| = 3$.

Distance between 3 and 0
is $|3-0| = |3| = 3$.
OR $|0-3| = |-3| = 3$.

$|a|$ is 0 or a positive integer.

If $a < 0$, then $-a$ is a positive number.

EX. $|-2| = -(-2) = +2 = 2 > 0$

For any number a, the absolute value of a is denoted by

$$|a| = \begin{cases} a, & a \geq 0 \\ -a, & a < 0 \end{cases}$$

$2 < 3 \Rightarrow |2| < |3|$

$= 2 \qquad = 3$

$-3 < -2 \Rightarrow |-3| > |-2|$

$= -(-3) = 3 \qquad = -(-2) = 2$

Note: ① *Negative integer* $< 0 <$ *Positive integer*

② *For any positive integers a and b,* $a < b \Rightarrow |a| < |b|$

③ *For any negative integers a and b,* $a < b \Rightarrow |a| > |b|$

6. Exponents

(1) For any positive integer a,

$$a^n > 0, \text{ whether } n \text{ is even or odd}$$

For example, $2^2 = 4 > 0, \ 2^3 = 8 > 0$

(2) For any negative integer a,

$$a^n > 0, \text{ if } n \text{ is even}$$
$$a^n < 0, \text{ if } n \text{ is odd}$$

For example, $(-2)^2 = (-2)(-2) = +4 > 0, \ (-2)^3 = (-2)(-2)(-2) = -8 < 0$

Note: For any number $a(\neq 0)$ and positive integers m, n

① Same base product rule:

$$a^m a^n = a^{m+n}$$

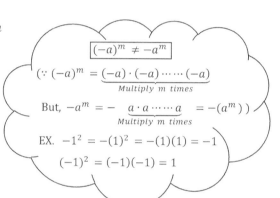

$$(-a)^m \neq -a^m$$

$$(\because (-a)^m = \underbrace{(-a) \cdot (-a) \cdots\cdots (-a)}_{Multiply \ m \ times}$$

$$\text{But, } -a^m = - \underbrace{a \cdot a \cdots\cdots a}_{Multiply \ m \ times} = -(a^m))$$

EX. $-1^2 = -(1)^2 = -(1)(1) = -1$

$(-1)^2 = (-1)(-1) = 1$

(Keep the base and add the powers)

For example, $a^2 a^3 = (a \cdot a)(a \cdot a \cdot a) = a^5 = a^{2+3}$

② Same base quotient rule:

$$\frac{a^m}{a^n} = a^{m-n}$$

(Keep the base and subtract the denominator's power from the numerator's power)

For example, $\dfrac{a^3}{a^2} = \dfrac{a \cdot a \cdot a}{a \cdot a} = a = a^1 = a^{3-2}, \quad \dfrac{a^2}{a^3} = \dfrac{a \cdot a}{a \cdot a \cdot a} = \dfrac{1}{a} = a^{-1} = a^{2-3}$

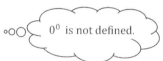

0^0 is not defined.

$$③ \quad a^0 = 1 \ ; \ 1^0 = 1, \ 2^0 = 1, \ 3^0 = 1, \cdots\cdots$$

(No matter what value a has, any nonzero number a raised to the power of 0 is 1.)

For example, $2 + a^0 = 2 + 1 = 3, \quad 2 - a^0 = 2 - 1 = 1$

$$2 \cdot a^0 = 2 \cdot 1 = 2, \quad \frac{2}{a^0} = \frac{2}{1} = 2$$

2-2   The Rational Numbers

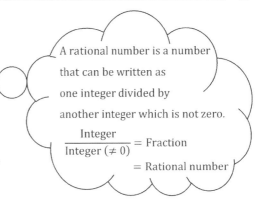

A rational number is a number that can be written as one integer divided by another integer which is not zero.

$$\frac{\text{Integer}}{\text{Integer}\,(\neq 0)} = \text{Fraction}$$

$$= \text{Rational number}$$

1. Classification of Rational Numbers

A *rational number* is a fraction whose numerator and denominator are both integers, and is defined by $\frac{a}{b}$, for any integers a and b ($\neq 0$).

In the fraction $\frac{a}{b}$, b ($\neq 0$), a is called the *numerator*, b is called the *denominator*, and the horizontal bar is called the *fraction bar*.

Note : In a fraction, $\frac{a}{0}$ (zero denominator) is undefined.

Rational numbers
- Integers
  - Positive integers (1, 2, 3, $\cdots$)
  - Zero (0)
  - Negative integers ($-1, -2, -3, \cdots$)
- Non – integers : $\frac{1}{2}, \frac{1}{3}, -\frac{1}{2}, -\frac{1}{3},\ 0.3 = \frac{3}{10},\ -1.5 = -\frac{15}{10}, \cdots$
  (Fractions)

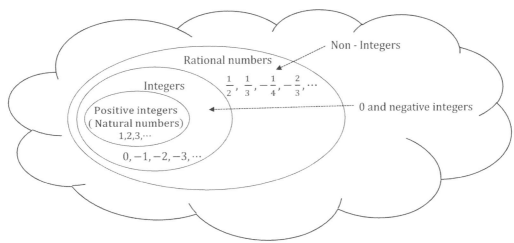

2. Notation of Rational Numbers

(1) Proper Fraction: $\frac{a}{b}$, $a < b(\neq 0)$

The numerator is less than the denominator.

For example, $\frac{1}{2}, \frac{1}{3}, \frac{2}{3}, \frac{1}{4}, \frac{2}{4}\left(=\frac{1}{2}\right), \frac{3}{4}, \cdots\cdots$

Keep dividing by common factors until no common factors (other than 1) appear.
OR
Divide by GCF in one step.

$$\underset{\div 2\quad \div 3}{\overset{\div 6}{\frac{12}{18} = \frac{6}{9} = \frac{2}{3}}}\qquad \underset{\div 2\quad \div 2}{\overset{\div 4}{\frac{16}{20} = \frac{8}{10} = \frac{4}{5}}}$$

Note : ① *A proper fraction is in lowest term*

if the numerator and denominator have no common factor except 1.

② *Reduce fractions to lowest terms.*

Divide both the numerator and denominator by the greatest common factor, GCF.

For example, $\dfrac{2}{4} = \dfrac{1}{2}$ $(\because \text{GCF }(2, 4) = 2)$,

$\dfrac{12}{18} = \dfrac{2}{3}$ $(\because \text{GCF }(12, 18) = 6)$, $\dfrac{16}{20} = \dfrac{4}{5}$ $(\because \text{GCF }(16, 20) = 4)$

(2) Improper Fraction: $\dfrac{a}{b}$, $a \geq b (\neq 0)$

The numerator is greater than or equal to the denominator.

For example, $\dfrac{3}{2}$, $\dfrac{3}{3} (= 1)$, $\dfrac{4}{3}$, $\dfrac{5}{3}$, $\dfrac{6}{3} (= 2)$, $\cdots\cdots$

Note: $0 = \dfrac{0}{a}$, *for any* $a(\neq 0)$

$1 = \dfrac{1}{1} = \dfrac{2}{2} = \dfrac{3}{3} = \cdots\cdots$, $2 = \dfrac{2}{1} = \dfrac{4}{2} = \dfrac{6}{3} = \cdots\cdots$, $3 = \dfrac{3}{1} = \dfrac{6}{2} = \dfrac{9}{3} = \cdots\cdots$, *and so on.*

A whole number can be written by an improper fraction.

(3) Mixed Number: $a\dfrac{b}{c} = a + \dfrac{b}{c}$, $(c \neq 0)$

A whole number plus a fraction.

For example, $1\dfrac{1}{2}$, $2\dfrac{1}{3}$, $2\dfrac{2}{3}$, $\cdots\cdots$

Note:

(1) *A mixed number is converted into an improper fraction.*

① $\quad a\dfrac{b}{c} = a + \dfrac{b}{c} = \dfrac{ac}{c} + \dfrac{b}{c} = \dfrac{ac+b}{c}$

For example, $2\dfrac{3}{4} = (2) + (\dfrac{3}{4}) = \dfrac{8}{4} + \dfrac{3}{4} = \dfrac{11}{4}$

$-2\dfrac{3}{4} = -\left(2\dfrac{3}{4}\right) = -\dfrac{11}{4}$

$-2\dfrac{3}{4} \neq (-2) + (\dfrac{3}{4})$

A fraction represents a part of a whole number.

-2 is not a part of a whole number.

$-2\dfrac{3}{4} = -\left(2\dfrac{3}{4}\right) = -\left(2 + \dfrac{3}{4}\right) = -2 - \dfrac{3}{4}$

② $\quad a\dfrac{b}{c} = \dfrac{(a \cdot c)+b}{c}$

Step 1. Multiply the whole number, a, and the denominator, c,

and then add this product to the numerator, b .

Step 2. Divide by the denominator, c .

For example, $2\dfrac{3}{4} = \dfrac{(2 \cdot 4)+3}{4} = \dfrac{8+3}{4} = \dfrac{11}{4}$

(2) *An improper fraction is converted into a mixed number.*

To convert an improper fraction $\dfrac{a}{b}$, *divide a by b.*

Let q be the quotient and r be the remainder. Then, $a = q \cdot b + r$

Since a mixed number $q\dfrac{r}{b}$ is equal to $\dfrac{q \cdot b + r}{b}$, $q\dfrac{r}{b} = \dfrac{q \cdot b + r}{b} = \dfrac{a}{b}$

Now, the improper fraction $\dfrac{a}{b}$ is converted into a mixed number $q\dfrac{r}{b}$.

For example, $\dfrac{3}{2} = 1\dfrac{1}{2}$, $\dfrac{14}{3} = 4\dfrac{2}{3}$, $\dfrac{10}{4} = 2\dfrac{2}{4} = 2\dfrac{1}{2}$, $\cdots$

Convert   *Reduce*

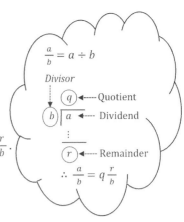

3. Magnitude of Rational Numbers

To compare fractions $\dfrac{a}{b}$ $b(\neq 0)$ and $\dfrac{c}{d}$ $d(\neq 0)$, the denominators b and d must be the same.

So, we use the least common multiple (LCM) of the denominators which is called the *least common denominator* (LCD).

Example Compare $\dfrac{3}{4}$ and $\dfrac{2}{5}$.

$\dfrac{a}{b} = \dfrac{c}{d}$, $b(\neq 0)$, $d(\neq 0)$
$\Longleftrightarrow ad = bc$ (Cross Product)

Since their cross products are not the same ($4 \cdot 2 \neq 3 \cdot 5$), $\dfrac{3}{4} \neq \dfrac{2}{5}$.

Since LCD = LCM(4, 5) = 20, find the equal ratios with a denominator of 20.

Then, $\dfrac{3}{4} = \dfrac{15}{20}$ and $\dfrac{2}{5} = \dfrac{8}{20}$.

Now we can compare them. Since $15 > 8$, $\dfrac{15}{20} > \dfrac{8}{20}$. Therefore, $\dfrac{3}{4} > \dfrac{2}{5}$.

Note: Equivalent fractions are different fractions with the equal value.

For example, $\dfrac{1}{2}, \dfrac{2}{4}, \dfrac{3}{6}, \dfrac{4}{8}$, and $\dfrac{5}{10}$ are all equivalent fractions.

4. Operations of Rational Numbers

(1) Adding and Subtracting Fractions

For any integers $a, b, c(\neq 0)$, and $d(\neq 0)$,

1) If the denominators are the same:

$$\frac{a}{c} + \frac{b}{c} = \frac{a+b}{c}, \qquad \frac{a}{c} - \frac{b}{c} = \frac{a-b}{c}$$

Just add or subtract the numerators of the fractions.

The denominator will not be changed. Simplify the result if needed.

For example, $\dfrac{3}{5} + \dfrac{1}{5} = \dfrac{3+1}{5} = \dfrac{4}{5}$, $\qquad \dfrac{3}{5} - \dfrac{1}{5} = \dfrac{3-1}{5} = \dfrac{2}{5}$

2) If the denominators are different:

$$\frac{a}{c} + \frac{b}{d} = \frac{ad}{cd} + \frac{bc}{cd} = \frac{ad+bc}{cd}, \quad \frac{a}{c} - \frac{b}{d} = \frac{ad}{cd} - \frac{bc}{cd} = \frac{ad-bc}{cd}$$

Rewrite all fractions as equivalent fractions with common denominators.

That is, ① Identify the least common denominator of the fractions by multiplying both
numerator and denominator of each fraction by the same number.

② Keep the denominator and add or subtract the numerators.

③ Simplify the result if needed.

For example, $\dfrac{2}{3} + \dfrac{1}{4} = \dfrac{8}{12} + \dfrac{3}{12} = \dfrac{8+3}{12} = \dfrac{11}{12}, \quad \dfrac{2}{3} - \dfrac{1}{4} = \dfrac{8}{12} - \dfrac{3}{12} = \dfrac{8-3}{12} = \dfrac{5}{12}$

Note: $\quad \dfrac{a}{b} = \dfrac{a \cdot c}{b \cdot c} = \dfrac{a}{b} \cdot \dfrac{c}{c} = \dfrac{a}{b} \cdot 1 = \dfrac{a}{b}$

Multiplying the numerator and denominator of a fraction by a non-zero number c is multiplying

a fraction by $\dfrac{c}{c}$, *because* $\dfrac{c}{c} = 1$ *and multiplying any number by 1 does not affect the value of the*

original number. Thus, that results in many ratios.

For example,

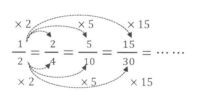

We can find an equal ratio by multiplying or
dividing the numerator and denominator of a
fraction by a non-zero number.

Example

Simplify the expression: $\dfrac{1}{2} + \dfrac{2}{3} - \dfrac{3}{4}$

To rewrite the fraction using the least common denominator (LCD),

keep the expressions separate and get them to have the same LCD.

$$\begin{array}{r} 2\,)\,\underline{2 \quad 3 \quad 4} \\ 1 \quad 3 \quad 2 \end{array}$$

$\therefore$ LCM $= 2 \cdot 1 \cdot 3 \cdot 2 = 12$

Therefore, $\dfrac{1}{2} + \dfrac{2}{3} - \dfrac{3}{4} = \dfrac{1 \cdot 6}{2 \cdot 6} + \dfrac{2 \cdot 4}{3 \cdot 4} - \dfrac{3 \cdot 3}{4 \cdot 3} = \dfrac{6}{12} + \dfrac{8}{12} - \dfrac{9}{12} = \dfrac{6+8-9}{12} = \dfrac{5}{12}$

(2) Adding and Subtracting Mixed Numbers

1) Using improper fraction

If fractions are mixed numbers, convert the fractions into improper fractions

using $a\dfrac{b}{c} = \dfrac{(a \cdot c) + b}{c}$, $c \neq 0$, and then identify the LCD of the fractions.

Example $2\dfrac{1}{3} + 1\dfrac{1}{4} = \dfrac{7}{3} + \dfrac{5}{4} = \dfrac{7 \cdot 4}{12} + \dfrac{5 \cdot 3}{12} = \dfrac{28+15}{12} = \dfrac{43}{12} = 3\dfrac{7}{12}$

   Similarly, $2\dfrac{1}{3} - 1\dfrac{1}{4} = \dfrac{7}{3} - \dfrac{5}{4} = \dfrac{7 \cdot 4}{12} - \dfrac{5 \cdot 3}{12} = \dfrac{28-15}{12} = \dfrac{13}{12} = 1\dfrac{1}{12}$

2) Using the concept, $a\dfrac{b}{c} = a + \dfrac{b}{c}$

Separate the mixed number into the whole number part and the fraction part.

① Adding

Add the fraction parts of the mixed numbers. Find a common denominator before adding, if needed. If the result is an improper fraction, rewrite as a mixed number. Then, add the whole number parts of the mixed numbers. Finally, combine the result of the fraction part with the result of the whole number part.

Example Add $3\dfrac{3}{4}$ and $5\dfrac{5}{6}$.

$\dfrac{3}{4} + \dfrac{5}{6} = \dfrac{9}{12} + \dfrac{10}{12} = \dfrac{19}{12} = 1\dfrac{7}{12}$ and $3 + 5 = 8$

$\therefore\ 3\dfrac{3}{4} + 5\dfrac{5}{6} = (3+5) + \left(\dfrac{3}{4} + \dfrac{5}{6}\right) = 8 + 1\dfrac{7}{12} = 9\dfrac{7}{12}$

$3\dfrac{3}{4} = 3\dfrac{6}{8} = 3\dfrac{9}{12} = 3\dfrac{12}{16} = \cdots$

② Subtracting

Subtract the fraction parts of the mixed numbers. You may need to find a common denominator or to borrow before you do the subtraction. Reduce the fraction, if needed. Then, subtract the whole number parts of the mixed numbers. Finally, combine both results.

Example Subtract $3\dfrac{3}{4}$ and $5\dfrac{5}{6}$.

Since $\dfrac{3}{4} = \dfrac{9}{12}$, $\dfrac{5}{6} = \dfrac{10}{12}$ and $\dfrac{9}{12} < \dfrac{10}{12}$, we have to borrow.

$-a\dfrac{b}{c} \neq -a + \dfrac{b}{c}$

$-a\dfrac{b}{c} = -\left(a\dfrac{b}{c}\right) = -\left(a + \dfrac{b}{c}\right)$

$= -a - \dfrac{b}{c}$

Since $3\dfrac{3}{4} = 3\dfrac{9}{12} = 2\dfrac{21}{12}$,

$3\dfrac{3}{4} - 5\dfrac{5}{6} = 2\dfrac{21}{12} - 5\dfrac{10}{12} = (2-5) + \left(\dfrac{21}{12} - \dfrac{10}{12}\right) = -3 + \dfrac{11}{12} = -\left(3 - \dfrac{11}{12}\right)$

$= -\left(2\dfrac{12}{12} - \dfrac{11}{12}\right) = -\left((2-0) + \left(\dfrac{12}{12} - \dfrac{11}{12}\right)\right) = -\left(2 + \dfrac{1}{12}\right) = -2\dfrac{1}{12}$

Note : Whole numbers are written as fractions with a denominator of 1.

So you can simplify expressions containing whole numbers in order to add or subtract.

For example, $3 - \dfrac{4}{5} = \dfrac{3}{1} - \dfrac{4}{5} = \dfrac{15}{5} - \dfrac{4}{5} = \dfrac{15-4}{5} = \dfrac{11}{5} = 2\dfrac{1}{5}$

(3) Multiplying and Dividing Fractions

1) Multiplying Fractions

For any integers a, b, $c(\neq 0)$, and $d(\neq 0)$,

$$\boxed{\dfrac{a}{c} \times \dfrac{b}{d} = \dfrac{ab}{cd}}$$

> For any integers $a, b, c(\neq 0)$, and $d(\neq 0)$,
>
> $$\dfrac{a}{c} \times \dfrac{b}{d} = \dfrac{ab}{cd}$$
>
> Multiply the numerators and then divide by the product of the denominators.

① Multiply the numerators of the two fractions. Multiply the denominators of the two fractions.

No LCD is required for this. If possible, reduce the fractions to the simplest form before multiplying.

For example, $\dfrac{3}{5} \times \dfrac{6}{8} = \dfrac{3}{5} \times \dfrac{3}{4} = \dfrac{3 \times 3}{5 \times 4} = \dfrac{9}{20}$

Reduce   multiply

> Same signs : $\begin{cases} (+) \cdot (+) = (+) \\ (-) \cdot (-) = (+) \end{cases}$
>
> Different signs : $\begin{cases} (+) \cdot (-) = (-) \\ (-) \cdot (+) = (-) \end{cases}$

Note : A mixed number is in simplest form if the mixed number has fractional part in lowest term.

For example,   if $a = 3\dfrac{6}{8}$, then $a = 3\dfrac{3}{4}$ is the simplest form.

②   If the fractions are mixed numbers, convert both mixed numbers into improper fractions before multiplying . Then reduce the fraction to get the simplest form.

For example, $2\dfrac{1}{4} \times 3\dfrac{2}{3} = \dfrac{9}{4} \times \dfrac{11}{3} = \dfrac{33}{4} = 8\dfrac{1}{4}$

③   To multiply fractions and whole numbers, consider the whole numbers as fractions with a denominator of 1.

For example, $2 \times 2\dfrac{3}{4} = \dfrac{2}{1} \times \dfrac{11}{4} = \dfrac{11}{2} = 2\dfrac{1}{2}$

$$a \times b = 1$$

$\Rightarrow$ a and b are reciprocals of each other.

For any $a(\neq 0)$, $\dfrac{0}{a} = 0$; $a \times 0 = 0$

$\therefore$ 0 has no reciprocal.

For any $a(\neq 0)$, $\dfrac{a}{0}$ is undefined.

2) Dividing Fractions

For any nonzero rational numbers $a, b, c,$ and d,

$$\frac{a}{c} \div \frac{b}{d} = \frac{a}{c} \times \frac{d}{b} = \frac{ad}{cb}$$

To divide two fractions, convert division into multiplication using the reciprocal of the fraction you are dividing by (also known as second fraction). If possible, reduce the fractions to lowest terms before multiplying, and then calculate the product.

For example, $\quad \dfrac{3}{5} \div \dfrac{6}{7} = \dfrac{3}{5} \times \dfrac{7}{6} = \dfrac{1 \times 7}{5 \times 2} = \dfrac{7}{10}$

(4) Multiplying and Dividing Mixed Numbers

Rewrite the mixed numbers as improper fractions and then multiply or divide the fractions. Simplify the result if needed.

For example, $\quad 3\dfrac{1}{2} \times 2\dfrac{4}{5} = \dfrac{7}{2} \times \dfrac{14}{5} = \dfrac{7 \times 7}{1 \times 5} = \dfrac{49}{5} = 9\dfrac{4}{5}$

$\qquad\qquad\qquad 3\dfrac{1}{2} \div 2\dfrac{4}{5} = \dfrac{7}{2} \div \dfrac{14}{5} = \dfrac{7}{2} \times \dfrac{5}{14} = \dfrac{1 \times 5}{2 \times 2} = \dfrac{5}{4} = 1\dfrac{1}{4}$

(5) Complex Fraction

For any nonzero rational numbers $a, b, c,$ and d,

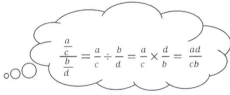

$$\frac{\dfrac{a}{c}}{\dfrac{b}{d}} = \frac{a}{c} \div \frac{b}{d} = \frac{a}{c} \times \frac{d}{b} = \frac{ad}{cb}$$

$$\frac{a}{c} \div \frac{b}{d} = \frac{\dfrac{a}{c}}{\dfrac{b}{d}} = \frac{ad}{cb}$$

$\leftarrow$ Outer product
$\leftarrow$ Inner product

Multiply the inner numbers c and b and consider it as the denominator.
Multiply the outer numbers a and d and consider it as the numerator.

If possible, reduce the numerators or denominators in a complex function.

EX. $\dfrac{3}{5} \div \dfrac{6}{7} = \dfrac{1}{5} \div \dfrac{2}{7} = \dfrac{\frac{1}{5}}{\frac{2}{7}} = \dfrac{7}{10}$

reduce 3 and 6 (numerators)

EX. $\dfrac{4}{9} \div \dfrac{16}{21} = \dfrac{1}{9} \div \dfrac{4}{21} = \dfrac{1}{3} \div \dfrac{4}{7} = \begin{cases} \dfrac{\frac{1}{3}}{\frac{4}{7}} = \dfrac{7}{12} \text{ or} \\ \dfrac{1}{3} \times \dfrac{7}{4} = \dfrac{7}{12} \end{cases}$

reduce 4 and 16 (numerators)
reduce 9 and 21 (denominators)

For example, $\quad \dfrac{3}{5} \div \dfrac{6}{7} = \dfrac{\frac{3}{5}}{\frac{6}{7}} = \dfrac{3 \times 7}{5 \times 6} = \dfrac{21}{30} = \dfrac{7}{10}$

$\qquad\qquad\qquad 4\dfrac{1}{6} \div 3 = \dfrac{4\frac{1}{6}}{\frac{3}{1}} = \dfrac{\frac{25}{6}}{\frac{3}{1}} = \dfrac{25}{18} = 1\dfrac{7}{18}$

$\qquad\qquad\qquad 5 \div \dfrac{2}{3} = \dfrac{\frac{5}{1}}{\frac{2}{3}} = \dfrac{\frac{5}{1}}{\frac{2}{3}} = \dfrac{15}{2} = 7\dfrac{1}{2}$

Exercises

#1   Express the following as an integer:

(1) An increase of 30%

(3) 1 week later

(2) A loss of 5 points

(4) 3 degrees below 0

#2   Draw a number line for each point.

$A(+2),\ B(-2),\ C(-5),\ D(+3),\ E(0),\ F(-1)$

#3   Order the integers from least to greatest.

$+4,\ -5,\ 0,\ -8,\ +5,\ +1$

#4   State the distance between the two points.

(1) $+4$ and $+2$

(3) -2 and -5

(2) -1 and $+2$

(4) -1 and 0

#5   For $a = 2,\ b = -3,\ c = -4$, find the value for the following:

(1) $a + b$

(6) $b - c$

(11) $a \times b \times c$

(2) $b + c$

(7) $a \div b$

(12) $b \div a \times c$

(3) $a \times b$

(8) $b \div c$

(13) $b \times c \div a$

(4) $b \times c$

(9) $a + b + c$

(5) $a - b$

(10) $a - b - c$

#6   Find the value of a in the following:

(1) $2 - 3 - a = -2$

(3) $-1 - (-a) - (+2) = -3$

(2) $-a - (-2) + (-3) = 4$

(4) $2 + (-a) - (-5) = -1$

(5) $(-1) + (-1)^2 + (-1)^3 + (-1)^4 = a$　　　　(8) $a \times 3 = -9$

(6) $1 - 2 + 3 - 4 + 5 - 6 + 7 - 8 = a$　　　　(9) $a \div (-4) = 3$

(7) $-2 \times a = 4$　　　　(10) $9 \div -a = -3$

(11) $a = b + c$, where $\begin{aligned} b &= -1^n + (-1)^{2n} - (-1)^{3n}, \quad n \text{ is odd} \\ c &= -1^n + (-1)^{2n} - (-1)^{3n}, \quad n \text{ is even} \end{aligned}$

#7　a is a negative integer. Determine whether the value is positive or negative for the following:

(1) $-a$　　　　(4) $(-a)^2$　　　　(7) $-a^3$

(2) $-(-a)$　　　　(5) $-a^2$　　　　(8) $-a^0$

(3) a^2　　　　(6) $(-a)^3$

#8　For any integers $a, b,$ and $c,$ find the value which satisfies the following conditions:

(1) $b + c$ when $a = 3,$ $ab = 5,$ and $ac = 7$

(2) ac when $ab = 2$ and $a(b + c) = 10$

(3) bc when $ac = 9$ and $c(a - b) = -5$

#9　Simplify each expression.

(1) $-4 + 3^2 \div (-3)$　　　　(4) $(-2)^3 + (-1)^5 - (-1)^4 \times (-2)^3$

(2) $-25 - ((-3)^3 \div 3) \times (-5)$　　　　(5) $80 + [\,3 - \{5 \times (-2)^3 - 7\,\}\,] \div 5 \times (-2)^3$

(3) $2 \times ((-5)^2 + 3) - 12 \div (-3)$

#10　In each of the following, determine whether the statement is true or false.

(1) Rational numbers are integers.

(2) The smallest natural number is zero.

(3) All natural numbers are integers.

(4) All natural numbers are rational numbers.

(5) Zero is a rational number.

(6) Rational numbers are positive integers and negative integers.

(7) There is an integer between two different integers.

(8) All integers are rational numbers.

(9) The smallest integer is -1.

(10) There are many rational numbers between two different rational numbers.

#11 Order the following numbers from least to greatest:

(1) $-7, \dfrac{1}{2}, 0, \dfrac{1}{3}, -\dfrac{1}{4}$

(2) $2, -5, -2, \dfrac{1}{4}, \dfrac{3}{5}$

(3) $\dfrac{3}{4}, -\dfrac{1}{2}, \dfrac{7}{12}, \dfrac{5}{6}, -\dfrac{1}{5}$

(4) $-\dfrac{2}{3}, \dfrac{4}{6}, -\dfrac{1}{6}, -\dfrac{3}{4}, \dfrac{5}{8}$

#12 Determine whether the following fractions are equal or not:

(1) $\dfrac{3}{2}$ and $\dfrac{5}{8}$

(2) $\dfrac{3}{4}$ and $\dfrac{4}{7}$

(3) $\dfrac{0}{3}$ and $\dfrac{0}{5}$

(4) $\dfrac{6}{5}$ and $\dfrac{3}{2}$

(5) $\dfrac{2}{3}$ and $\dfrac{10}{15}$

(6) $\dfrac{4}{9}$ and $\dfrac{16}{36}$

(7) $\dfrac{3}{7}$ and $\dfrac{14}{28}$

(8) 3 and $\dfrac{12}{4}$

(9) $\dfrac{5}{24}$ and $\dfrac{4}{16}$

(10) $\dfrac{5}{12}$ and $\dfrac{7}{16}$

#13 Convert each mixed number into an improper fraction.

(1) $3\dfrac{5}{6}$

(2) $-4\dfrac{2}{5}$

(3) $5\dfrac{3}{8}$

(4) $-6\dfrac{8}{12}$

(5) $7\dfrac{6}{15}$

(6) $-2\dfrac{12}{18}$

(7) $4\dfrac{15}{45}$

(8) $3\dfrac{16}{28}$

(9) $-5\dfrac{2}{8}$

(10) $2\dfrac{16}{20}$

#14 Two different integers a and b are in-between $-5\dfrac{2}{3}$ and $-\dfrac{1}{2}$.

Find the greatest number of $a - b$ and the smallest number of $a + b$.

#15 Find all the fractions with a denominator of 20 in lowest terms between $-\dfrac{3}{4}$ and $\dfrac{3}{5}$.

#16 Solve the following expressions in lowest terms:

(1) $3\frac{2}{5} + 2\frac{1}{4}$

(2) $7\frac{3}{4} - 3\frac{1}{3}$

(3) $4\frac{2}{8} - 3\frac{2}{7}$

(4) $5\frac{3}{8} + 2\frac{4}{6}$

(5) $2\frac{4}{9} - \frac{3}{5}$

(6) $3\frac{3}{6} + 2\frac{3}{5}$

(7) $-\frac{2}{5} + \left(-1\frac{1}{3}\right) - \left(-1\frac{2}{5}\right)$

(8) $2\frac{1}{3} - \left(+\frac{5}{2}\right) + \frac{1}{3} - \left(-\frac{3}{4}\right)$

#17 Write each answer as a fraction. If possible, convert it to a mixed number in lowest term.

(1) $2\frac{1}{3} \times 3\frac{3}{5}$

(2) $3\frac{2}{7} \times 2$

(3) $2\frac{1}{5} \times \frac{1}{6}$

(4) $1\frac{1}{4} \times 2\frac{2}{3}$

(5) $4 \times 2\frac{3}{5}$

(6) $\frac{2}{5} \times 1\frac{1}{3}$

(7) $\frac{2}{3} \times 2\frac{1}{5}$

(8) $3\frac{1}{2} \div \frac{1}{3}$

(9) $1\frac{3}{4} \div \frac{3}{5}$

(10) $3\frac{2}{3} \div 2$

(11) $1\frac{3}{5} \div 2\frac{2}{5}$

(12) $3 \div 2\frac{1}{2}$

(13) $2\frac{3}{4} \div 3\frac{4}{6}$

#18 For any rational numbers a and b, find the value of $a + b$ in lowest terms for the following expressions:

(1) $a = \frac{2}{3} + \left(-\frac{1}{2}\right)$ and $b = -\frac{1}{4} - \left(+\frac{2}{3}\right)$

(2) $a = \frac{3}{4} - \left(-\frac{1}{5}\right)$ and $b = -\frac{3}{2} + \left(-\frac{4}{5}\right)$

(3) $a = -5\frac{1}{2} + \left(-3\frac{2}{3}\right)$ and $b = 2\frac{1}{3} - \left(-4\frac{3}{4}\right)$

(4) $\frac{3}{4} - \left(+\frac{1}{3}\right) + \left(-\frac{1}{2}\right) - \left(-\frac{5}{6}\right) = \frac{a}{b}$

(5) $\left(-2\frac{1}{3}\right) + (-2) - \left(+1\frac{1}{4}\right) - \left(-1\frac{3}{2}\right) = -\frac{a}{b}$

(6) $a = \frac{3}{2} \times \left(-\frac{22}{8}\right)$ and $b = -\frac{7}{9} \times \left(-\frac{3}{14}\right)$

(7) $a = 2\frac{1}{3} \times \left(3\frac{1}{2}\right)$ and $b = -3 \times \left(-\frac{2}{9}\right)$

(8) $a = -2^3 \times \left(-\frac{2}{3}\right)^2$ and $b = \left(-\frac{1}{2}\right)^2 \times (-1^2)$

(9) $a = (-1)^{100} \times \left(-\frac{1}{2}\right)^3$ and $b = \left(-\frac{1}{3}\right)^2 \times \frac{3}{4}$

(10) $\frac{3}{4} \times a = 1$ and $b \times \left(-2\frac{2}{3}\right) = 1$

(11) $a = -\frac{6}{4} \div -\frac{3}{2}$ and $b = 2\frac{1}{4} \div -3\frac{2}{3}$

(12) $a = -\frac{1}{8} \div (-\frac{1}{4})^2$ and $b = \frac{2}{3} \div -4$

(13) $\left(-\frac{1}{3}\right) \div \left(-\frac{5}{12}\right) \div \left(-\frac{2}{3}\right)^2 = \frac{a}{b}$

(14) $a \div (-4) = -3$ and $b = -3\frac{1}{2} \div \frac{3}{4}$

(15) $-\frac{3}{4} \times a = 1$ and $a \times (-b) = \frac{8}{3}$

(16) $(-2)^3 \div \left(-\frac{1}{4}\right) \times (-1^2) \div (-\frac{4}{3}) = \frac{a}{b}$

(17) $a = -5\frac{1}{2} \times 2\frac{1}{3} \times -\frac{4}{7}$ and $b = (7 - 5\frac{1}{6}) \times \frac{3}{2} \div (-8\frac{1}{4})$

(18) $a = (-2)^3 \div \left(-\frac{2}{5}\right) \div \left(1\frac{1}{4}\right)$ and $b = 2\frac{3}{4} \div \left(-3\frac{2}{3}\right) \div -1^3$

(19) $a = (-2)^3 \div \left(-\frac{3}{2}\right) \div \frac{4}{3}$ and $b = (-1)^5 \div \frac{2}{3} \times (-\frac{9}{6})$

(20) $a = 3 \circ (2\frac{1}{3} \circ \frac{7}{9})$ and $b = 2\frac{1}{3} \odot (\frac{1}{3} \odot \frac{1}{4})$,

   where $m \circ n = m \div n$ and $m \odot n = \frac{mn}{m+n}$ for any rational numbers m and n.

#19 For any number a, $-1 < a < 0$. Order the following expressions from greatest to least:

$$\frac{1}{a}, \ -\frac{1}{a}, \ -\frac{1}{a^2}, \ -a, \ (-\frac{1}{a})^2, \ a^2$$

#20 Simplify the following expressions:

(1) $(-1)^3 - (-1^2) + (-1)^{100} - (-1)^{99} \times (-1)^5 + (-1)^0$

(2) $4\frac{1}{2} - 6 \times \left\{ \frac{3}{4} \div \frac{9}{8} - \frac{1}{3} \times \left(-\frac{1}{2}\right)^3 \right\}$

(3) $-2 + \left[-3^2 \div 3 \times \frac{1}{4} - \left\{ \frac{1}{2} - \left(-\frac{1}{3}\right)^2 \right\} \times (-2) \div \frac{1}{3} \right]$

(4) $3 - \left[2 - \left\{ (-2)^3 \times (-3^2) - (-1)^3 \div \frac{1}{2} \right\} - 2 \right]$

(5) $2 - \dfrac{2}{2 - \dfrac{1}{2 - \frac{1}{3}}}$

Chapter 3. Equations

3-1 Variables and Expressions

1. Definition

(1) Variable

A *variable* is a letter that represents an unknown number.

For example, $2x$ is the product of the number 2 and an unknown number named x.

(2) Expression

An *expression* is a number phrase without an equal sign or inequality sign.

(3) Term

A *term* is an expression of a single number or a product of numbers and variables.

For example, $3x + 5$ has two terms $3x$ and 5

(4) Constant

A *constant* is a term that doesn't contain any variable.

For example, 6 is the constant of the expression $4x + 5y + 6$

(5) Coefficient

A *coefficient* is a number that is multiplied by a variable.

For example, the coefficient of $8x$ is 8.

(6) Monomial expression

A *monomial expression* is an expression of only one term which is product of numbers and variables.

For example, 2, x, $3x$, $\frac{1}{2}xy$

(7) Polynomial Expression

A *polynomial expression* is an expression of two or more terms combined by addition and/or subtraction.

For example, $3x + 4$, $\frac{1}{3}xy + 2x - 5$

3-2 Equations

An equation has two expressions separated by an equal sign ($=$). The value of the expression on the left side of the equal sign must be the same as the value of the expression on the right side.

1. Definition

An expression using an equal sign between two expressions is called an *equation*.

For example, $5 + 6 = 11$ and $2x + 4 = 10$ are equations. $3 + 4 = 8$ is also an equation.

But $2x + 4$ is not an equation because the expression doesn't have an equal sign.

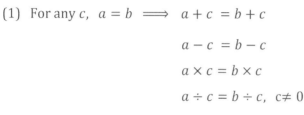

True equation

Depends on the value of x

False equation

2. Properties of Equations

(1) For any c, $a = b \implies a + c = b + c$

$$a - c = b - c$$
$$a \times c = b \times c$$
$$a \div c = b \div c, \quad c \neq 0$$

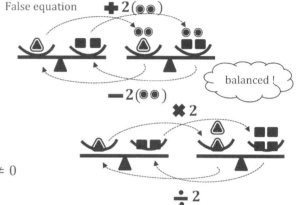

balanced !

The expressions on both sides of the equal sign must have the same value.

(This keeps the equation balanced.)

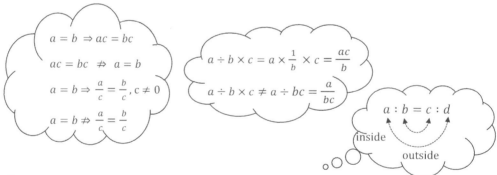

$$a = b \Rightarrow ac = bc$$
$$ac = bc \nRightarrow a = b$$
$$a = b \Rightarrow \frac{a}{c} = \frac{b}{c}, c \neq 0$$
$$a = b \nRightarrow \frac{a}{c} = \frac{b}{c}$$

$$a \div b \times c = a \times \frac{1}{b} \times c = \frac{ac}{b}$$
$$a \div b \times c \neq a \div bc = \frac{a}{bc}$$

$$a : b = c : d$$
inside
outside

(2) $\dfrac{a}{b} = \dfrac{c}{d}$ ($a : b = c : d$) $\implies$ 1) $ad = bc$ ($\because$ Cross product)

2) $\dfrac{a+b}{b} = \dfrac{c+d}{d}$ ($\because \dfrac{a}{b} = \dfrac{c}{d} \Rightarrow \dfrac{a}{b} + 1 = \dfrac{c}{d} + 1 \Rightarrow \dfrac{a+b}{b} = \dfrac{c+d}{d}$)

3) $\dfrac{a-b}{b} = \dfrac{c-d}{d}$ ($\because \dfrac{a}{b} = \dfrac{c}{d} \Rightarrow \dfrac{a}{b} - 1 = \dfrac{c}{d} - 1 \Rightarrow \dfrac{a-b}{b} = \dfrac{c-d}{d}$)

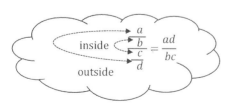

inside
outside
$$\frac{\frac{a}{b}}{\frac{c}{d}} = \frac{ad}{bc}$$

4) $\dfrac{a+b}{a-b} = \dfrac{c+d}{c-d}$ ($\because \dfrac{\frac{a+b}{b}}{\frac{a-b}{b}} = \dfrac{\frac{c+d}{d}}{\frac{c-d}{d}} \Rightarrow \dfrac{a+b}{a-b} = \dfrac{c+d}{c-d}$)

5) $\dfrac{a}{b} = \dfrac{c}{d} = \dfrac{e}{f} = \dfrac{a+c+e}{b+d+f}$, $b + d + f \neq 0$

($\because$ let $\dfrac{a}{b} = \dfrac{c}{d} = \dfrac{e}{f} = k$. Then $a = bk$, $c = dk$, $e = fk$.

$\therefore a + c + e = (b + d + f)k$ $\quad \therefore k = \dfrac{a+c+e}{b+d+f}$, $b + d + f \neq 0$)

3. Linear Equations and their Solutions

(1) Linear Equations

A *linear equation* is an equation with one variable which has the highest power of 1.

For example, the equation $ax = b$,

where a and b are constants, is a linear equation with a variable x.

(2) Solutions

The *solution* (*root*) of an equation is a number which makes the equation a true expression.

For example, the linear equation $2x = 10$ is true when $x = 5$.

Thus, $x = 5$ is the solution of the equation $2x = 10$.

For the linear equation $ax = b$, where a and b are constants, the solution is

1) $a \neq 0 \Rightarrow x = \dfrac{b}{a}$ (Only one solution)

2) $a = 0 \Rightarrow$ ① $b \neq 0 \Rightarrow$ No solution ($\because$ Division is not defined.)

② $b = 0 \Rightarrow$ All real numbers

4. Solving Linear Equations

> For $ax = b$,
>
> $0 \cdot x = 0 \implies x = \dfrac{0}{0}$ is the unlimited solution.
>
> $0 \cdot x = b\,(\neq 0) \implies x = \dfrac{b}{0}$ is not defined. There is no solution.

(1) Steps for Solving Equations with One Variable

Step 1: If the coefficient of the equation is a fraction or decimal, change it to an integer by multiplying a proper number by both sides of the equation.

Step 2: If there are parentheses, remove them by using the distributive property.

Step 3: Convert it into a simpler equation; transfer all variables to one side of the equation and transfer all numbers to the other side of the equation in the form of $ax = b$, $a \neq 0$.

Step 4: Solve for the variable x by dividing each side of the equation ($ax = b$) by a.

This gives $x = \dfrac{b}{a}$, $a \neq 0$. Therefore, $\dfrac{b}{a}$ is the one and only solution of $ax = b$.

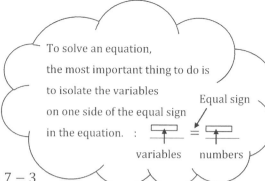

To solve an equation, the most important thing to do is to isolate the variables on one side of the equal sign in the equation. :

Equal sign

variables   numbers

Example

$$2x + 3 = 7 \xLongrightarrow[\text{Subtract 3 from both sides}]{} 2x + 3 - 3 = 7 - 3$$

$$\xLongrightarrow[\text{variable=number}]{} 2x = 4$$

$$\xLongrightarrow[\text{Multiply each side by } \frac{1}{2} \text{ or divide each side by 2}]{} 2x \times \frac{1}{2} = 4 \times \frac{1}{2} \quad \text{or} \quad \frac{2x}{2} = \frac{4}{2}$$

$$\xLongrightarrow[\text{reduce to lowest terms}]{} x = 2$$

Example

$$\frac{5}{2}x + 4 = x - 2 \xLongrightarrow[\text{Multiply each side by 2}]{} 5x + 8 = 2x - 4$$

$$\xLongrightarrow[\text{transfer (Step 3)}]{} 5x - 2x = -4 - 8$$

$$\xLongrightarrow[\text{simplify}]{} 3x = -12$$

$$\xLongrightarrow[\text{Divide each side by 3 or multiply each side by } \frac{1}{2}]{} \frac{3x}{3} = \frac{-12}{3} \quad \text{or} \quad 3x \times \frac{1}{3} = -12 \times \frac{1}{3}$$

$$\xLongrightarrow[\text{simplify}]{} x = -4$$

(2) Steps for Solving Word Problems

1) Assign a variable to represent the unknown number.

2) Find an equation for the problem.

3) Solve the equation.

4) Check the solution by substituting the solution into the variable in the equation.

Solution Problems:	Distance, Rate, and Time Problems:
The salt refers to as the solute, the water is the solvent, and the resulting mixture as the solution (water+salt). The amount (the concentration) of salt in the solution expresses how salty the salt water is.	**Distance = Rate × Time** $$\text{Rate} = \frac{\text{Distance}}{\text{Time}}, \quad \text{Time} = \frac{\text{Distance}}{\text{Rate}}$$
$$\textbf{Concentration} = \frac{\textbf{The amount of salt (solute)}}{\textbf{The total amount of solution}}$$ Concentration is normally expressed as a percent (%), multiplied by 100 .	If the rate is in miles per hour, then the distance must be in miles and the time in hours. If the time is in minutes, convert it to hours (dividing by 60) to find the distance in miles.
The amount of salt (solute) **= (Concentration) × (The total amount of solution)**	*Match the units!!*

5. Absolute Values

The absolute value of a number a is denoted by $|a|$.

If $a \geq 0$, then $|a| = a$

If $a < 0$, then $|a|$ is the corresponding positive number.

> The absolute value of a number a, denoted by $|a|$ is the distance of a from zero on a number line. Since a distance is always a positive number, absolute value is never negative. For example,
> $|2| = 2$
> : two units to the right of zero on a number line.
> $|-2| = 2$
> : two units to the left of zero on a number line.

Example

$$|0| = 0, \ |1| = 1, \ |2| = 2, \ |-1| = 1, \ |-2| = 2$$

If the number has no minus sign in front of it, the absolute value is not changed.

But if the number has minus sign in front of it, we remove the minus sign to find the absolute

Value. For example,

if $a = -1$, then $-a = -(-1) = 1$ and if $a = -2$, then $-a = -(-2) = 2$.

Therefore, we define the absolute value of a as

$$|a| = \begin{cases} a & \text{if } a \geq 0 \\ -a & \text{if } a < 0 \end{cases}$$

> $|-2| = -(-2) = 2$
> $-|-2| = -2$

6. Linear Equations with Absolute Values

The absolute value of a real number a, denoted by $|a|$, is defined as follows:

$$|a| = \begin{cases} a & \text{if } a \geq 0 \\ -a & \text{if } a < 0 \end{cases}$$

Example $\quad |x - 1| = 2x + 3$

$$\Longrightarrow \begin{cases} x - 1 \geq 0 \ (x \geq 1) \ \Rightarrow \ |x-1| = x - 1 = 2x + 3 \ ; \ x = -4 \ ; \ \text{not possible} \\ x - 1 < 0 \ (x < 1) \ \Rightarrow \ |x-1| = -(x-1) = 2x + 3 \ ; \ x = -\dfrac{2}{3} \end{cases}$$

$$\therefore \ x = -\frac{2}{3}$$

Example $\quad |x + 1| + |x + 2| = 5$

Since $(x + 1 = 0 \Rightarrow x = -1)$ and $(x + 2 = 0 \Rightarrow x = -2)$,

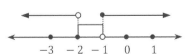

Consider $x < -2, \ -2 \leq x < -1, \ x \geq -1$

$$\Longrightarrow \begin{cases} ① \ x < -2 \ ; \ -(x+1) - (x+2) = 5 \ ; \ x = -4 \\ ② \ -2 \leq x < -1 \ ; -(x+1) + (x+2) = 5 \ ; \ 0 \cdot x = 4 \ ; \ \text{no solution} \\ ③ \ x \geq -1 \ ; \ x + 1 + x + 2 = 5 \ ; \ x = 1 \end{cases}$$

$\therefore \ x = -4, \ x = 1$

Example $|x + 1| = |x + 3|$

$\implies x + 1 = \pm (x + 3)$

$|a| = |b| \Rightarrow a = \pm b$

$$\implies \begin{cases} x + 1 = x + 3 \implies 0 \cdot x = 2 ; \ \text{no solution} \\ x + 1 = -(x + 3) \implies 2x = -4 ; x = -2 \end{cases}$$

$\therefore \ x = -2$

Exercises

 # 1. Find the value for each expression.

(1) $x + 3$ if $x = -3$

(2) $-(x + 1)$ if $x = -5$

(3) $x^3 - 2x - 5$ if $x = -1$

(4) $2xy - 4$ if $x = -3$, $y = 2$

(5) $x^2 - 5y$ if $x = -2$, $y = -3$

(6) $\dfrac{2}{x} + \dfrac{3}{y}$ if $x = -\dfrac{1}{4}$, $y = -\dfrac{1}{6}$

(7) $\dfrac{y}{x} - \dfrac{x}{y}$ if $x = 2$, $y = -3$

(8) $x^{99} - x^6$ if $x = -1$

(9) $\dfrac{3}{a} - 2b^2$ if $a = \dfrac{1}{4}$, $b = -3$

(10) $\dfrac{1}{a} - \dfrac{2}{b} - \dfrac{3}{c}$ if $a = -\dfrac{1}{4}$, $b = -\dfrac{1}{2}$, $c = -6$

(11) $2(a - b) - (a^2 - b^2)$ if $a = -2$, $b = 3$

(12) $|a - 2| - |3ab - a|$ if $a = -1$, $b = 2$

 # 2. Simplify each expression.

(1) $\dfrac{1}{2}x \cdot (-6)$

(2) $-\dfrac{2}{3}(6x - 9)$

(3) $(6x - 2) \div \left(-\dfrac{3}{10}\right)$

(4) $\dfrac{1}{2}(4x - 6) - \dfrac{2}{3}\left(9x - \dfrac{3}{4}\right)$

(5) $-3x + 6x - 2x - 5$

(6) $2x + 4y - \{3x - (5 - 2y)\} - 3$

(7) $(2a - 3b) - (5a - 2b) - (4 - a)$

(8) $3m^2 - (5m - m^2 - 1) + 2m$

(9) $\dfrac{3t^3 - 4t^2}{2t}$

(10) $\dfrac{2a^2b - 3ab^2 - 5ab}{ab}$

(11) $(3x - 9) \div \dfrac{3}{2} - 8\left(\dfrac{3}{4}x - 2\right)$

(12) $\dfrac{x - 2}{3} - \dfrac{2x - 1}{4} - \dfrac{3 - x}{2}$

 # 3. Find an expression for the perimeter.

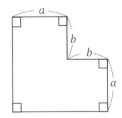

4. Find an expression for the shaded area.

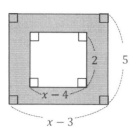

5. Which expression is different from the others?

(1) $x \div (y \times z)$

(2) $x \div y \div z$

(3) $x \times \dfrac{1}{y} \div z$

(4) $x \div (y \div z)$

(5) $x \times \dfrac{1}{y} \times \dfrac{1}{z}$

6. Find $a + b$ and $a - b$ when a is a coefficient of x, and b is a constant for the expression:

$\dfrac{4x-3}{2} - \dfrac{2x-1}{3}$

7. The coefficient of x is 3 and the constant is 5 for the form $2x + b - (ax + 3)$.

Find $a \cdot b$ and $\dfrac{a}{b}$

8. For two expressions A and B, if you add $2x + 3$ to A then you get $5x + 7$ and if you subtract $-3x - 4$ from B then you get $2x + 5$. What is $2A - 3B$?

9. Solve the following equations for x:

(1) $3x - 2 = 7$

(2) $2x + 3 = 3x - 2$

(3) $5x - 2 = \dfrac{1}{2}x - 1\dfrac{1}{4}$

(4) $0.2x - 0.3 = 0.4x - 0.5$

(5) $\dfrac{3}{4}\left(x - \dfrac{1}{3}\right) = \dfrac{1}{2}\left(\dfrac{1}{5} + 4x\right)$

(6) $\dfrac{x-3}{2} - 1 = \dfrac{x}{4} - 3$

(7) $3(1 - 2x) + 7 = -2x - 2$

(8) $3 - \dfrac{2x-1}{3} = 5x - \dfrac{x-2}{6}$

\# 10. Find a constant a that makes the equation $4(a + x) = 2(2x + 3) + 6$ true for all values of x.

\# 11. For any positive integers $a, b,$ and c, a is divided by b and the remainder is c.

Express the quotient using $a, b,$ and c.

\# 12. For any non-zero constants $a, b(b \neq 1),$ and $c,$ find the value of $\frac{1}{abc}$ such that $a + \frac{1}{b} = 1$ and

$b + \frac{1}{c} = 1.$

\# 13. $a = \frac{2}{3}, b = \frac{3}{4},$ and $c = -\frac{4}{5}.$ Find the value of $\frac{ab+bc+ca}{abc}.$

\# 14. Find the sum of all possible solutions for the equation $|2x - 3| = 5$.

\# 15. $A = 2x - 3,$ $B = 3x + 4.$ The ratio of A to B is $3 : 5$.

When a is the solution of x, find the value of $-\frac{a}{3} + 3$.

\# 16. For any constants $a, b,$ the solution of the equation $3x - 2 = ax - 4$ is $x = -1$ and

the solution of the equation $\frac{1}{2}x + b = ax + 3$ is $x = -2$. Find $a \cdot b$

\# 17. The solution of the equation $2ax + 5 = -3$ is half of the solution of the equation

$x - 5 = 3x + 7$. Find the value of $3a - 4$.

\# 18. For any constants a and b, $\frac{1}{a} - \frac{1}{b} = 3$ $(ab \neq 0)$. Find the value of $\frac{5a-3ab-5b}{a-b}.$

19. $\begin{cases} (1) \ \dfrac{a+3}{4} - \dfrac{2x-2}{3} = 1 \\ (2) \ \dfrac{3a-2}{2} - \dfrac{2a-x}{3} = 1 \end{cases}$

When the ratio of the solution of (1) to the solution of (2) is $1 : 4$, find the value of a.

20. For any x, the equation $3x - 5a = 2bx + 6$, where a and b are constants, is always true.

Find the value of $\dfrac{a}{2b}$.

21. The solution of an equation $\dfrac{2x-5a}{3} + x + 4 = 8$ is a negative integer.

Find the greatest value of a.

22. $a@b = ab^2 + a^2b$ When $\dfrac{1}{a} = 2$, $\dfrac{1}{b} = -3$, find the value of $b@a$.

23. How much water should be added to 30 ounces of a 20% salt solution to produce a 15% solution?

24. Richard has 20 ounces of a 15% of salt solution.

How much salt should he add to make it a 20% solution?

25. Richard drives to place A at 30 miles per hour. 20 minutes after he departs, Nichole goes to the place A at 50 miles per hour. How long will it take until Richard meets Nichole?

26. Richard wants to make 50 ounces of a 10% salt solution by mixing a 7% salt solution with a 15% salt solution. How many ounces of a 7% salt solution must be mixed?

27. Richard spends two-thirds of the money in his pocket to buy a book. He now has 4 dollars left. How much money did he have at the beginning?

28. A bag is on sale for a 15% discount. Nichole paid $60, including a 6% sales tax. What was the original price of the bag (rounded to the nearest hundredth)?

29. Richard's aunt is 51 years old. She is three times as old as the sum of the ages of Richard and his sister. Richard is 7 years younger than his sister. How old is Richard's sister?

30. The sum of three consecutive odd integers is 153. Find the biggest number of these three integers.

31. The tens digit of a certain two-digit integer is 3. If the digits of the number are interchanged, the number will be 1 less than two times the original number. Find the original number.

32. Richard is 5 years old and Nichole is 12 years old. In how many years will Nichole be two times Richard's age?

33. Richard takes 3 hours to finish a job if he works alone. Nichole takes 2 hours to finish the same job if she works alone. How long will it take them to finish the job if they work together?

34. Nichole took 8 days to finish a job and Richard took 6 days to finish the same job. If Nichole worked $3\frac{1}{3}$ days alone and then Nichole and Richard worked together to finish the job, how many days did they work together?

35. Nichole checked a book out from a library. She read $\frac{1}{3}$ of the book on the first day, $\frac{1}{4}$ of the book on the second day, and 39 pages on the third day. She now has to read $\frac{1}{5}$ of the book to finish. How many pages does the book have?

36. Richard finishes a job alone in 5 hours. If Nichole helps him, they can finish the job together in 1 hour 40 minutes. How many hours would it take Nichole to work alone to finish the job?

37. Nichole goes out to eat at a restaurant. Her total bill is $23, including a 15% tip. How much was the dinner?

38. A movie ticket price for children is $3 less than the adult ticket price. Nichole paid $36 for 2 adults and 3 children. What is the price of an adult ticket?

39. Nichole and Richard live in the same home. They drove to a park to meet some friends. They started from their home at the same time. Nichole drove at 40 miles per hour and Richard drove at 50 miles per hour. Nichole arrived at the park 10 minutes late while Richard arrived 5 minutes early for their appointment. Find the distance from Richard and Nichole's home to the park.

40. Find the value for each of the following:

(1) $|4|$ (3) $-|-3|$ (5) $|-7-5|$ (7) $|3|-|8|$

(2) $|-5|$ (4) $|2-6|$ (6) $|5|+|-5|$ (8) $|-9|+(-9)$

41. Solve the following equations:

(1) $|x-3| = 5x + 2$ (2) $|x-4| + |x+2| = 10$ (3) $|x-2| - |5-x| = 0$

Chapter 4. Inequalities

4-1 Inequalities with One Variable

1. Algebraic Inequality Symbols

$<$: Less than

$>$: Greater than

$\leq$: Less than or equal to

$\geq$: Greater than or equal to

$\neq$: Not equal to

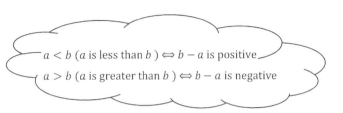

$a < b$ (a is less than b) $\Leftrightarrow b - a$ is positive

$a > b$ (a is greater than b) $\Leftrightarrow b - a$ is negative

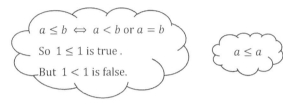

$a \leq b \Leftrightarrow a < b$ or $a = b$

So $1 \leq 1$ is true.

But $1 < 1$ is false.

$a \leq a$

2. Definition

Inequality is an expression using the algebraic inequality symbols to show the relationship between the values of numbers or variables.

Example

(1) $x > 2$ (x is greater than 2) means:

2 is not included (open circle) in the solution.

(2) $-2 \leq x < 1$ (x is greater than or equal to -2 and less than 1) means:

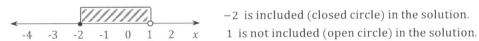

-2 is included (closed circle) in the solution.

1 is not included (open circle) in the solution.

3. Properties of Inequalities

For any real numbers a, b and c, the following properties apply:

(1) Transitive Property:

$$a < b , \; b < c \;\Rightarrow\; a < c$$

(2) Adding or subtracting the same number to or from each side of an inequality does not change the direction of the inequality symbol.

$$a < b \;\Rightarrow\; a + c < b + c \quad \text{and} \quad a - c < b - c$$

(3) Multiplying or dividing each side of an inequality by the same positive number does not change the direction of the inequality symbol.

$$a < b,\ c > 0 \ \Rightarrow\ a \cdot c < b \cdot c \quad \text{and} \quad \frac{a}{c} < \frac{b}{c}$$

$a < b,\ c = 0 \ \Rightarrow\ a \cdot c = b \cdot c$

$a < b \qquad\qquad \not\Rightarrow\ a^2 < b^2\ (\because\ -2 < -1,\ \text{but}\ (-2)^2 > (-1)^2)$

$0 < a < b \qquad \Rightarrow\ a^2 < b^2$

$a < b \qquad\qquad \Rightarrow\ a^3 < b^3\ (\because\ -3 < -2,\ \text{and}\ (-3)^3 < (-2)^3)$

(4) Multiplying or dividing both sides of an inequality by the same negative number will change the direction by reversing the inequality symbol).

$$a < b,\ c < 0 \ \Rightarrow\ a \cdot c > b \cdot c \quad \text{and} \quad \frac{a}{c} > \frac{b}{c}$$

$-2x < 1 \xRightarrow{\times\left(-\frac{1}{2}\right)} x > -\frac{1}{2}$

$a < x \leq b \xRightarrow{\times(-1)} -a > -x \geq -b;\ -b \leq -x < -a$

(5) $\boxed{\begin{array}{l} a < b \iff a - b < 0 \\ a > b \iff a - b > 0 \end{array}}$

(6) $\boxed{\begin{array}{l} \text{If } a \text{ and } b \text{ have the same sign} \Rightarrow a \cdot b > 0,\ \dfrac{a}{b} > 0,\ \text{and}\ \dfrac{b}{a} > 0 \\[2mm] \text{If } a \text{ and } b \text{ have different signs} \Rightarrow a \cdot b < 0,\ \dfrac{a}{b} < 0,\ \text{and}\ \dfrac{b}{a} < 0 \end{array}}$

$a < b \iff -a > -b$

(7) Expanded Properties

$$\begin{array}{l} a < x < b \qquad\qquad \Rightarrow\ a + c < x + c < b + c,\quad a - c < x - c < b - c \\[2mm] a < x < b,\ c > 0 \ \Rightarrow\ ac < xc < bc,\quad \dfrac{a}{c} < \dfrac{x}{c} < \dfrac{b}{c} \\[2mm] a < x < b,\ c < 0 \ \Rightarrow\ ac > xc > bc,\quad \dfrac{a}{c} > \dfrac{x}{c} > \dfrac{b}{c} \end{array}$$

4. Solutions of Linear Inequalities

To solve a linear inequality with one variable (to identify the values of x which satisfy the inequality), isolate the variable on one side of the inequality and solve it exactly like a linear equation. The solution will consist of intervals or unions of intervals.

(1) The linear Inequality is formed by

$$ax < b\ (ax \leq b) \quad \text{or} \quad ax > b\ (ax \geq b) \quad \text{for any}\ a \neq 0$$

(2) The solution of the inequality $ax < b$ is

  ① $x < \dfrac{b}{a}$, when $a > 0$ (Positive a)

  ② $x > \dfrac{b}{a}$, when $a < 0$ (Negative a)

  ③ All real numbers, when $a = 0$ and $b > 0$

  ④ Not defined (no solution), when $a = 0$ and $b \leq 0$

For $ax < b$,

if $a = 0$ then $0 \cdot x < b$

$\Rightarrow \begin{cases} ① \text{True , when } b > 0 \\ \quad \therefore \text{ All real numbers are the solution.} \\ ② \text{False, when } b \leq 0 \\ \quad \therefore \text{ Solution does not exist (no solution).} \end{cases}$

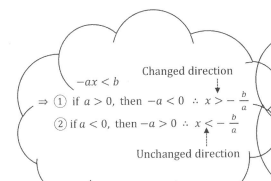

$-ax < b$

$\Rightarrow$ ① if $a > 0$, then $-a < 0$ $\therefore$ $x > -\dfrac{b}{a}$

  ② if $a < 0$, then $-a > 0$ $\therefore$ $x < -\dfrac{b}{a}$

Changed direction

Unchanged direction

Find k.

(1) The solution of $ax < b$ is $x < k$.

  $\Rightarrow$ Since the direction of the inequality symbol of the solution is unchanged, $a > 0$.

  From $ax < b$, $x < \dfrac{b}{a}$ $\therefore k = \dfrac{b}{a}$.

(2) The solution of $ax < b$ is $x > k$.

  $\Rightarrow$ Since the direction of the inequality symbol of the solution is changed, $a < 0$.

  From $ax < b$, $x > \dfrac{b}{a}$ $\therefore k = \dfrac{b}{a}$

5. Steps for Solving Word Problems

Step 1. Remove parentheses, using the distributive property.

Step 2. If there are fractions or decimals for coefficients of variables, change all the coefficients to integers by multiplying a proper number to both sides of the inequality symbol.

Step 3. Isolate the variables and numbers on each side of the inequality symbol.

Step 4. Simplify both sides of the symbol:

  For any $a(\neq 0)$, $ax < b$ ($ax \leq b$) or $ax > b$ ($ax \geq b$)

Step 5. Find the solution .

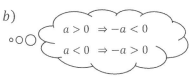

$a > 0 \Rightarrow -a < 0$

$a < 0 \Rightarrow -a > 0$

4-2   Graphing Linear Inequalities with One Variable

1. Solutions on a Number Line

For the linear inequality $ax < b$ or $ax \leq b$ with $a > 0$, the solution is given by the simpler inequality: $x < \dfrac{b}{a}$ or $x \leq \dfrac{b}{a}$.

This is because the solution has an infinite number of values and cannot be expressed in a simpler form. The usual representation of this solution is given by its graph on a number line, using open points and closed points.

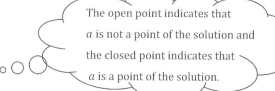

The open point indicates that a is not a point of the solution and the closed point indicates that a is a point of the solution.

(1) $x < a$

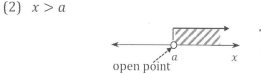

open point

The solution includes an infinite number of values less than a.

(2) $x > a$

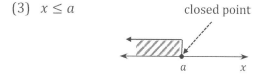

open point

The solution includes an infinite number of values greater than a.

(3) $x \leq a$

closed point

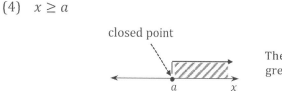

The solution includes an infinite number of values less than or equal to a.

(4) $x \geq a$

closed point

The solution includes an infinite number of values greater than or equal to a.

2. Compound (Combined) Inequalities (Double Inequalities)

A compound inequality consists of two inequalities joined by "and" or "or".

For $m < n$,

(1) $m < x < n \Rightarrow m < x$ and $x < n$

(2) $x > n$ or $x \leq m$

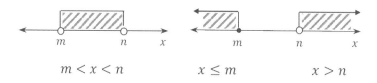

$m < x < n$ $\qquad$ $x \leq m$ $\qquad$ $x > n$

To solve this compound inequality, isolate the variable between the inequality symbols, or isolate the variable in each inequality.

Example

(1) $1 < 2x - 3 \leq 5 \Rightarrow 1 + 3 < 2x - 3 + 3 \leq 5 + 3$

$$\Rightarrow 4 < 2x \leq 8$$

$$\Rightarrow \frac{4}{2} < x \leq \frac{8}{2}$$

$$\Rightarrow 2 < x \leq 4 \Rightarrow x \leq 4 \ \text{and} \ x > 2$$

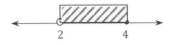

2 is not in the graph by placing an open circle above it.
4 is a point of the graph by filling in the circle above it.

(2) $3x + 4 \leq 2 + 2x \ \text{or} \ -2x < x + 3$

$$\Rightarrow \ x \leq -2 \ \text{or} \ 3x > -3$$

$$\Rightarrow \ x \leq -2 \ \text{or} \ x > -1$$

3. Linear Inequalities with Absolute Values

$$|a| = \begin{cases} a, & a \geq 0 \\ -a, & a < 0 \end{cases}$$

For $a > 0$, $|x| > a \Rightarrow x > a \ \text{or} \ -x > a$

Since $-x > a$ is equivalent to $x < -a$, $|x| > a$ is equivalent to $(x > a \ \text{or} \ x < -a)$.

Similarly, $|x + b| > a$ is equivalent to $(x + b > a \ \text{or} \ x + b < -a)$.

Therefore,

(1) $|x| < a \qquad \Leftrightarrow \ -a < x < a$

(2) $|x + b| < a \ \Leftrightarrow \ -a < x + b < a \ \text{or} \ -a - b < x < a - b$

(3) $|cx + b| < a \ \Leftrightarrow \ -a < cx + b < a \ \text{or} \ -a - b < cx < a - b$

$$\text{or} \ \frac{-a-b}{c} < x < \frac{a-b}{c} \ , \ c > 0$$

Example

(1) $2 < |x - 1| < 3$

① When $x - 1 \geq 0 \ (x \geq 1)$,

$2 < |x - 1| < 3 \Rightarrow 2 < x - 1 < 3 \Rightarrow 2 + 1 < x < 3 + 1 \Rightarrow 3 < x < 4$

So, $3 < x < 4$

② When $x - 1 < 0$ $(x < 1)$,

$2 < |x - 1| < 3 \Rightarrow 2 < -(x - 1) < 3 \Rightarrow 2 - 1 < -x < 3 - 1 \Rightarrow 1 < -x < 2$

$\Rightarrow -2 < x < -1$

So, $-2 < x < -1$

Therefore, $3 < x < 4$ or $-2 < x < -1$

(2) $|x + 2| + |x - 3| < 10$

Since $x + 2 = 0 \Rightarrow x = -2$ and $x - 3 = 0 \Rightarrow x = 3$,

consider the three cases, $x < -2$, $-2 \leq x < 3$, and $x \geq 3$.

① Case 1: When $x < -2$,

$|x + 2| + |x - 3| < 10 \Rightarrow -(x + 2) - (x - 3) < 10$

$\Rightarrow -2x < 9$

$\Rightarrow x > -\dfrac{9}{2}$

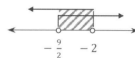

Since $x < -2$, $-\dfrac{9}{2} < x < -2$

② Case 2: When $-2 \leq x < 3$,

$|x + 2| + |x - 3| < 10 \Rightarrow (x + 2) - (x - 3) < 10$

$\Rightarrow 0 \cdot x < 5$; Always true

$\therefore -2 \leq x < 3$

③ Case 3: When $x \geq 3$,

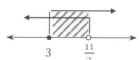

$|x + 2| + |x - 3| < 10 \Rightarrow (x + 2) + (x - 3) < 10$

$\Rightarrow 2x < 11$

$\Rightarrow x < \dfrac{11}{2}$

Since $x \geq 3$, $3 \leq x < \dfrac{11}{2}$

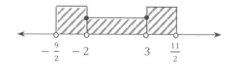

Therefore, the sum of all three intervals is $-\dfrac{9}{2} < x < \dfrac{11}{2}$.

Exercises

#1. Express each statement as an inequality.

(1) a is less than -3

(2) a is greater than or equal to 2

(3) a is greater than -1 and less than or equal to 1

(4) 3 more than twice a is greater than half of a

(5) 4 less than three time a is greater than or equal to a plus 2

(6) a is not greater than 0

#2. Solve the following inequalities:

(1) $x - 5 > 6$

(2) $x + 4 > 0$

(3) $6x > 3$

(4) $2x + 3 > 7$

(5) $3x - 4 > x + 3$

(6) $x + 5 > 3x$

(7) $-2x - 5 \leq 7$

(8) $-\frac{1}{3}x - 1 \leq 8$

(9) $-2x > 4$

(10) $-3x + 4 < -2x$

(11) $2x > 2(x + 3)$

(12) $5x - (7x - 6) \geq 3$

(13) $3x - (8x + 5) \leq 2$

(14) $2.5x - 1.5 > 3.5x + 4.5$

(15) $2(x + 1) - \frac{8x+1}{3} < 4$

(16) $\frac{4}{3}x - 4\left(\frac{1}{3}x + 2\right) > -1$

(17) $\frac{5x-3}{4} \geq x - \frac{5x+1}{3}$

(18) $0.3 - 0.2x < 0.4x - 0.1$

(19) $3 - 2ax < -3$ for $a < 0$

(20) $-ax - 1 \leq 2$ for $a < 0$

(21) $2ax > -a$ for $a < 0$

(22) $3x < 3x + 4$

#3. Solve the following inequalities. Then draw the solution on a number line:

(1) $2x - 4 > 4$

(3) $-\dfrac{x}{4} \geq 2$

(2) $-3x \leq \dfrac{x-1}{2} - 3$

(4) $0.3x - \dfrac{1+x}{2} < -\dfrac{2}{5}$

#4. Express the range of x as an inequality.

(1) Three times x minus 5 is greater than five times x plus 2.

(2) Two times the difference of x and 3 is less than or equal to three times the sum of $2x$ and 2.

#5. Express the range of x for the following expression when $-1 \leq x \leq 1$.

(1) $2x + 1$

(2) $-3x - 2$

(3) $\dfrac{1}{4}x - 3$

#6. Let $y = \dfrac{4 - 2x}{3}$.

(1) Find the range of y when $1 < x < 5$.

(2) Find the range of y when $-3 < x < -1$.

(3) Find the range of x when $2 \leq y \leq 4$.

#7. Find the sum of all positive integers which satisfy the following inequality:

$2(1 - x) + 6 \geq 3(x - 3) - 5$.

#8. The sum of three consecutive integers is greater than or equal to 69. Find the three integers with the smallest sum.

#9. How many positive integers satisfy the following inequalities?

(1) $3\left(\dfrac{1}{2}x - 1\right) < x + 2$

(2) $0.3(2 - x) \geq 0.1x - 0.2$

(3) $\dfrac{x+3}{2} - \dfrac{2-x}{3} < 1$

#10. Only 3 positive integers satisfy the inequality $3x - k \leq \frac{5-x}{2}$. Find the range of k.

#11. Find the constant k if:

(1) The inequality $\frac{1}{2}x - \frac{k}{3} < -1$ has the solution $x < 2$.

(2) The inequality $\frac{kx}{4} - \frac{1}{2} > 1$ has the solution $x < -1$.

(3) The inequality $\frac{2-kx}{5} - 2 \leq \frac{x}{2} + 1$ has the solution $x \leq -4$.

(4) Two inequalities $2(1 - 2x) - 3 \leq x - 5$ and $\frac{3k-2x}{3} \leq x + 2k$ have the same solution.

(5) The inequality $2 - kx < 2x + k$ has no solution.

(6) The inequality $-2kx + 5 > 6$ has the solution $x > 2$.

(7) $1 - 5x \leq 2x - 5k$ has -2 as a minimum value of the solution.

(8) The inequality $x - (3 + \frac{k}{2}) > 2x + k$ has no positive solution.

#12. The inequality $(-a + 2b)x + b - 3a \leq 0$ has the solution $x \leq -1$. Find the solution for the inequality $(a - b)x + a - 2b > 0$, where $b > 0$.

#13. Solve the following inequality for x and graph the solution:

(1) $|x - 2| \leq 0$

(2) $|3x + 9| > 0$

(3) $|x + 4| < 0$

(4) $|-2x + 1| + 3 \leq 6$

(5) $0 < |2 - 4x| < 8$

(6) $2 < |x + 1| < 3$

#14. At the store there is a bucketful of apples and peaches. An apple is worth 25¢ and a peach is worth 50¢. You want to buy 10 pieces of fruit. What is the maximum number of peaches you can buy with less than $10?

#15. Nichole wants to make a salt solution that is at most 10% salt by adding water to 50 ounces of a 15% salt solution. What is the amount of water she can add?

#16. Richard goes hiking in the mountain. He goes up the trail at a speed of 3 miles per hour and down the same trail at a speed of 4 miles per hour, while hiking for no longer than 2 hours. Find the maximum distance he can hike.

#17. Nichole plans to take a 3 miles walk in less than $\frac{1}{2}$ hour. She walks at a speed of 3 miles per hour at the beginning, then runs at a speed of 9 miles per hour for the rest. How far does she walk?

#18. Richard needs to produce a salt solution that is at least 12% salt after mixing 30 ounces of a 5% salt solution with a 15% salt solution. How many more ounces of a 15% salt solution must be needed?

#19. Nichole's last scores on three math tests were $93, 87,$ and 89 . When she takes her next test, she wants to have a total average of at least 92 points for all four tests. What score does she need to get on her next test?

#20. Richard is 14 years old and his dad is 48 years old. In how many years will his dad's age become less than twice Richard's age?

#21 . The lengths of three sides of a triangle are $x,$ $x + 2,$ and $x + 3,$ where x is a positive integer. Find the smallest length of the triangle.

#22. The shaded area is, at most, 66 square inches. Find the smallest integer for x.

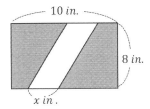

#23. There is a big sale going on at a book store. Nichole buys a book that is 20% off. But her $20 bill is not enough to buy it. What is the price range of the book?

#24. Suppose one boy can complete a task in 4 hours and one girl can complete the same task in 6 hours. A group of 5 boys and girls try to complete the task in 1 hour. Find the minimum number of boys to complete the task.

Chapter 5. Functions

5-1 Functions

1. Function

A *standard form* of a linear equation is formed by $ax + by = c$ for any constants $a \neq 0, b \neq 0,$ and $c.$ Solving the equation for y, we get the y in terms of x. For the relationship between x and y,

$y = ax + b, a \neq 0,$ y is called a *function* of x and is written $y = f(x)$

Each value of y is dependent on the chosen value of x, that is each value of x is assigned exactly one value of y. So, we call x the *independent variable* and y the *dependent variable*.

> A function is a relation between a set of inputs and a set of outputs that each input is related to only one output.
> If you plug in an x value in the FUNCTION, you will get out exactly one $f(x)$ value.

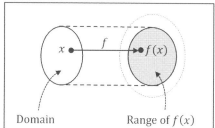

> Note: *A function is a relationship in which each element of the domain is paired with exactly one element of the range.*
>
> y *is a function of* x . $\underset{\text{represented by}}{\Longleftrightarrow}$ $f : x \longrightarrow y$ or $x \overset{f}{\longrightarrow} y$ or $y = f(x)$

2. Domain and Range

The *domain* of a function is the set of all input values for which $y = f(x)$ is defined.

The *range* of a function is the set of all output values for $y = f(x)$.

Example

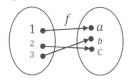

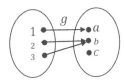

Domain : {1,2,3}

Range of $f(x)$: {a, b, c}

Domain : {1,2,3}

Range of $g(x)$: {a, b}

Domain   Range of $f(x)$

Note:

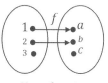

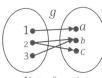

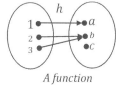

Not a function

(∵ $f(3)$ *does not exist.*)

Not a function

(∵ $g(2)$ *has 2 values.*)

A function

(∵ *Each value of* x *is assigned exactly one* $h(x)$.)

> y is a function of x if any value of y is determined by the value of x.

5-2 Graphing and Solving Functions

1. Coordinate Planes

(1) Coordinates

A *coordinate* is the number of a point assigned on a number line.

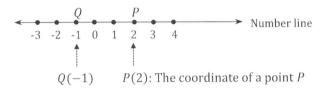

$Q(-1)$ $\qquad$ $P(2)$: The coordinate of a point P

Note : A number line is one-dimensional, and a coordinate plane is two-dimensional.

So, an equation with 1 variable is graphed on a number line and an equation with 2 different

variables is graphed on a coordinate plane.

(2) Coordinate Planes

When two number lines are perpendicular to each other at the zero points of the two lines,

1) their intersection is called the *origin*,

2) the horizontal number line is called the *x-axis*,

3) the vertical number line is called the *y-axis*,

4) the plane formed by the *x*-axis and the *y*-axis is called the *coordinate plane*.

Note: Each point P in the coordinate plane is assigned a pair of numbers.

(3) 4 Quadrants

4 Quadrants are four rectangular regions in a coordinate plane which are labeled with Roman numerals I, II, III , and IV. The coordinate plane is divided by the two axes into 4 quadrants, which are named in counterclockwise order.

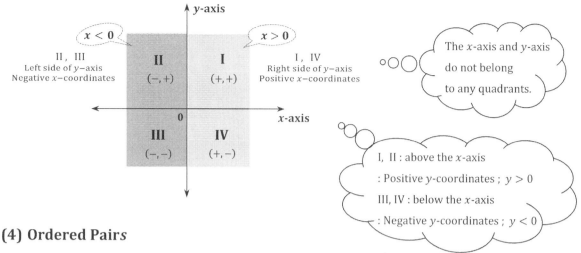

(4) Ordered Pairs

Coordinates, which are associated with points on the real line, can be introduced into the plane by ordered pairs of real numbers.

An ordered pair (a, b) of real numbers is the coordinates of the point P of which

1) a is the number of horizontal units moved from 0.

2) b is the number of vertical units moved from 0.

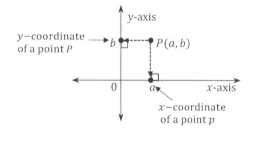

$(x, 0)$; The coordinate of a point on the x-axis
$(0, y)$; The coordinate of a point on the y-axis

Note: $(a, b) = (c, d)$ *if and only if* $a = c$ *and* $b = d$

$(a, b) \neq (b, a)$ *if* $a \neq b$

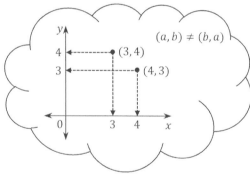

(5) Symmetric Transformation

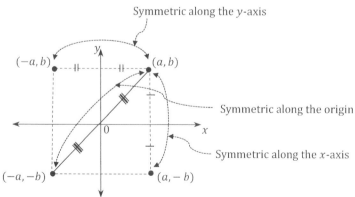

1) Symmetry of a point (a, b) along the x-axis ; $(a, -b)$: Opposite sign of y

2) Symmetry of a point (a, b) along the y-axis ; $(-a, \ b)$: Opposite sign of x

3) Symmetry of a point (a, b) along the origin ; $(-a, -b)$: Opposite signs of x and y

Note: (1) *The middle point between the two points $A(a_1, b_1)$ and $B(a_2, b_2)$ is*

$$\left(\frac{a_1 + a_2}{2}, \ \frac{b_1 + b_2}{2} \right)$$

(2) *The distance between the two points $A(a_1, b_1)$ and $B(a_2, b_2)$ is*

$$\sqrt{(a_2 - a_1)^2 + (b_2 - b_1)^2} \quad \textit{(By the Pythagorean Theorem)}$$

2. Graphing Functions

(1) Graphing $y = ax \ (a \neq 0)$

The graph of this function is known as a direct variation: A straight line through the origin (0,0).

1) $a > 0$

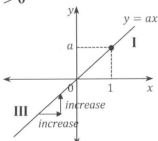

As the x values increase (rise) from left to right, the y values increase as well in the coordinate plane.

2) $a < 0$

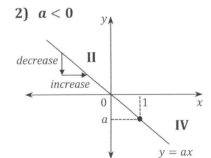

As the x values increase (rise) from left to right, the y values decrease(decline) in the coordinate plane.

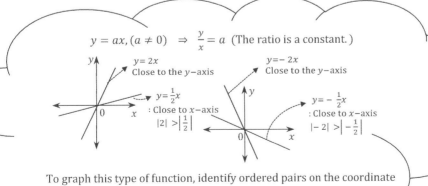

$y = ax, (a \neq 0) \ \Rightarrow \ \frac{y}{x} = a$ (The ratio is a constant.)

$y = 2x$
Close to the y-axis

$y = -2x$
Close to the y-axis

$y = \frac{1}{2}x$
: Close to x-axis
$|2| > \left| \frac{1}{2} \right|$

$y = -\frac{1}{2}x$
: Close to x-axis
$|-2| > \left| -\frac{1}{2} \right|$

To graph this type of function, identify ordered pairs on the coordinate plane and connect them with a straight line.

(2) Graphing $y = \dfrac{a}{x}$ $(a \neq 0)$

The graph of this function is known as an inverse variation: A pair of smooth curves which are symmetric along the origin $(0,0)$.

1) $a > 0$

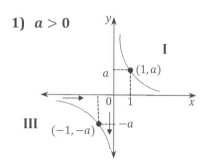

As the x values increase (rise)
from left to right,
the y values decrease
in the coordinate plane.

2) $a < 0$

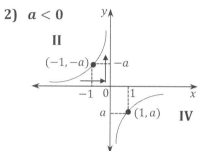

As the x values increase (rise)
from left to right,
the y values increase as well
in the coordinate plane.

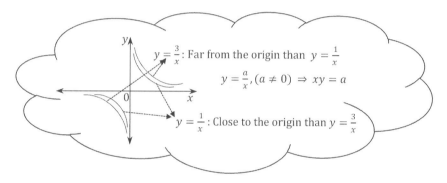

$y = \dfrac{3}{x}$: Far from the origin than $y = \dfrac{1}{x}$

$y = \dfrac{a}{x}, (a \neq 0) \Rightarrow xy = a$

$y = \dfrac{1}{x}$: Close to the origin than $y = \dfrac{3}{x}$

3. Steps for Solving Word Problems

(1) Determine the two variables x and y.

(2) Express the relationship between x and y as a function .

For example, $y = ax$ $(a \neq 0)$ or $y = \dfrac{a}{x}$ $(a \neq 0)$

(3) Find the solution for the equation.

(4) Check the solution.

Solution Problems:	Distance, Rate, and Time Problems:
The salt refers to as the solute, the water is the solvent, and the resulting mixture as the solution (water+salt). The amount (the concentration) of salt in the solution expresses how salty the salt water is.	**Distance = Rate × Time** $\text{Rate} = \dfrac{\text{Distance}}{\text{Time}}$, $\text{Time} = \dfrac{\text{Distance}}{\text{Rate}}$
$\text{Concentration} = \dfrac{\textbf{The amount of salt (solute)}}{\textbf{The total amount of solution}}$ Concentration is normally expressed as a percent (%), multiplied by 100.	If the rate is in miles per hour, then the distance must be in miles and the time in hours. If the time is in minutes, convert it to hours (dividing by 60) to find the distance in miles.
The amount of salt (solute) = **(Concentration)** × (**The total amount of solution**)	*Match the units!!*

Exercises

#1. The domain of a function $f(x) = -2x + 3$ is $\{0,\ 1,\ 2,\ 3\}$. Find the range of $f(x)$.

#2. The range of a function $f(x) = 2x$ is $\{-8,\ 0,\ 4,\ 8\}$. Find the domain of $f(x)$.

#3. Find the domain and range of the equation $y = |x| - 3$.

#4. The range of a function $g(x) = ax$ is $\{-2,\ 0,\ 2\}$ when $g(2) = -1$.

Find the domain of the function.

#5. $A = \{(-3,1), (-2,2), (-2,3), (-1,4), (0,5), (1,5)\}$ is the set of ordered pairs.

Is this relationship a function?

#6. For a function $f(x) = ax$, $f(3) = -4$. Find the value of $f(9)$.

#7. Find the value of $f(3) - f(2) + f(4)$ for the function $f(x) = \dfrac{3}{x}$.

#8. For the two functions $f(x) = ax + 2$ and $g(x) = \dfrac{b}{x} - 2$, $f(1) = g(-1) = 3$.

Find the value of $a + b$.

#9. For the two functions $f(x) = \dfrac{a}{x} + 2$ and $g(x) = -\dfrac{3}{x} + 5$, $3f(-2) = 2g(-3)$.

Find the value of b which satisfies $f(b) = g(b)$.

#10. Identify functions.

(1)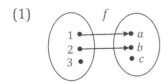

(2)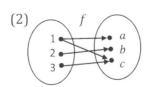

(3)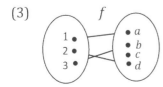

(4)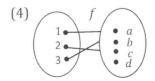

(5) 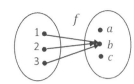   (6)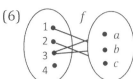

#11. For the function $f(3x - 2) = 2x - a$, $f(4) = 3$. Find the value of $f(1)$.

#12. For the two functions $f(x) = 2ax$ and $g(x) = \dfrac{2}{x} - 1$, $g(f(2)) = 3$. Find the value of a.

#13. Plot the following ordered pairs on the graph.

(1) $A(2, 3)$

(2) $B(-2, 3)$

(3) $C(2, -3)$

(4) $D(-5, 5)$

(5) $E(0, \ 5)$

(6) $F(4, 0)$

(7) $G(-3, 0)$

(8) $H(0, -7)$

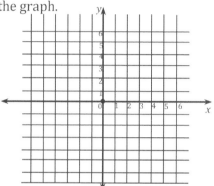

#14. Find the coordinates for each point on the graph.

(1) A

(2) B

(3) C

(4) D

(5) E

(6) F

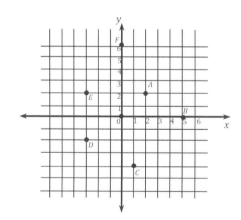

#15. Two points $P(a + 2, 4 - 2a)$ and $Q(2 - 2b, 3b + 1)$ are on the x-axis and y-axis respectively. Find the value of $a + b$.

#16. Find the length of the segment between the two points.

(1) $A(1, 2)$ and $B(1, -2)$ (3) $P(-3, 0)$ and $Q(5, 0)$

(2) $C(0, 3)$ and $D(-3, 3)$ (4) $S(-5, 0)$ and $T(-5, -6)$

#17. A point (a, b) is in the second quadrant of the coordinate plane.

Name the quadrant containing the following points:

(1) $(a, -b)$

(2) $(-b, a)$

(3) (b, a)

(4) $(-a, -b)$

(5) $(-a, b)$

(6) $(-b, -a)$

(7) (ab, a^2)

(8) $(-a, -ab)$

#18. Point B is reflected through the origin to point $A(3, 4)$.

Point C is obtained by reflecting point B across the y-axis. Find the area of a triangle $\triangle ABC$.

#19. Point $C(4, b)$ is the midpoint of Points $A(-2, 3)$ and $B(a, 9)$. Find the value of $a - b$.

#20. Which graphs are functions?

(1)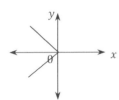

(2)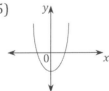

(3)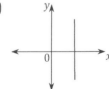

(4)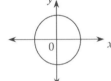

(5)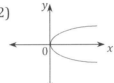

(6)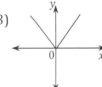

(7)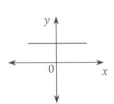

(8)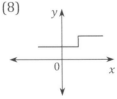

#21. Identify the function of the form $y = ax$ which passes through the origin and $(3, -4)$.

#22. Identify the functions of the form $y = ax$ or $y = \dfrac{a}{x}$ for the following graphs:

(1)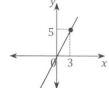

(2)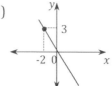

(3)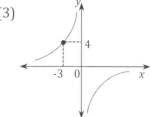

(4)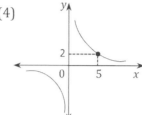

#23. Find the functions for the data in the tables below.

(1)

x	-4	-2	-1	1	2	4
y	-1	-2	-4	4	2	1

(2)

x	-2	-1	0	1	2	3
y	-1	1	3	5	7	9

(3)

x	1	2	3	4	6	12
y	12	6	4	3	2	1

(4)

x	-4	-2	0	2	4	6
y	2	1	0	-1	-2	-3

#24. The function $f(x) = -\dfrac{3}{2}x$ passes through a point $(a + 1, 2a - 3)$. Find the value of a.

#25. The function $y = ax$ passes through a point $(3, -15)$ and $(b, 10)$. Find the value of $a - b$.

#26. For any constants a and b, the function $f(x) = \dfrac{2a}{x}$ passes through the points

$(-2, 8)$ and $(4, b)$. Find the value of $a + b$.

#27. For any constants $a, b,$ and $c,$ the function $f(x) = \frac{a}{x}$ passes through the points $(b, 1)$, $(1, c)$,

and $(3, -1)$. Find the value of $a + b + c$.

#28. Two functions $f(x) = ax$ and $g(x) = \frac{b}{x}$ meet at the points $(3, 9)$ and $(-3, c)$.

Find the value of $a + b + c$.

#29. Two functions $y = -ax$ and $y = -\frac{2}{x}$ meet at Point $A(b, 8)$. Find the value of ab.

#30. Find the function of the form $y = ax$ or $y = \frac{a}{x}$ for the graph.

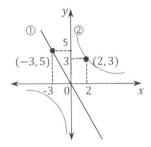

#31. The function $y = 3x$ passes through the two points, origin and A.

The area of the triangle $\triangle OAB$ is 54. Find the coordinate of Point A.

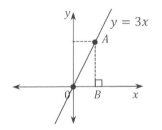

#32. Two points $P(3, a)$ and $Q(3, b)$ are on the graph $y = 3x$ and $y = -x,$ respectively.

Find the area of the triangle $\triangle OPQ$.

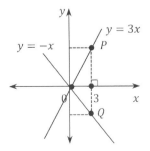

#33. Richard rides his bike from home to a park 5 miles away at a speed of x miles per hour for y hours. Find the relationship between x and y.

#34. Which one is not a function?

    (1) The sum of two variables x and y is 5.

    (2) The variable y is half of the variable x.

    (3) The perimeter (y inches) of a rectangle with one side of length (x inches).

    (4) 10 miles at a speed of x miles per hour for y minutes.

#35. A building needs to be painted. It takes 30 hours for 5 workers to finish the job. If the job has to be finished in 6 hours, how many workers are needed?

#36. Nichole wants to make a vegetable garden with an area of 200 square feet. Find the relation between the length (x feet) and width (y feet).

#37. The distance from A to B is 10 miles. Nichole drives at a speed of 35 miles per hour from A to B, and Richard drives at a speed of 25 miles per hour from B to A at the same time. How long will it take before they meet each other?

#38. Richard drives to a post office at a speed of 50 miles per hour. 5 minutes later, Nichole drives to the post office at a speed of 60 miles per hour. How long will it take before Nichole meets Richard?

#39. Richard drives to school at a speed of 40 miles per hour and returns back home at a speed of 30 miles per hour. Coming home, it takes him 10 more minutes than going school. How far is it from Richard's home to school?

#40. Nichole rides her bike halfway to school at 20 mph. She drives her car the rest of the way at 40 mph. Find Nichole's average speed to school.

#41. x ounces of a y% salt solution contains 3 ounces of salt. Find the relationship between x and y.

#42. Nichole wants to buy some books at a bookstore which are all the same price. If she buys 3 books, then she will be $2.50 short. If she buys 2 books, then she will have $5.00 left over. How much money does she have?

#43. 3 machines can do 5 jobs in 4 hours. How many hours will it take for 4 machines to do 6 jobs?

Chapter 6. Fractions and Other Algebraic Expressions

6-1 Decimals and Fractions

1. Decimals

A decimal consists of a whole number part and a decimal part (fraction part).

(1) Place Value

Every digit has a place value.

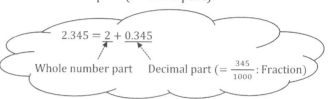

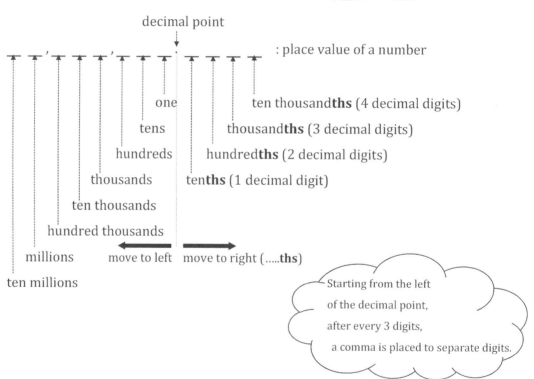

Note

① *The word "and" is used to separate the whole number part of the number from the decimal part.*

   So don't use the word "and" to represent a whole number.

② *To identify decimals, the decimal place value is written at the end of the number.*

   The decimal place values end in "ths".

③ *The hyphen "−" is used for numbers between 20 and 100.*

④ $2.4 = 2.40 = 2.400 = 2.4000 = \cdots\cdots$

   The zeroes placed at the end of a decimal have no effect on its value.

Example

24.5 : twenty-four and five tenths

("five" is a one digit number, and the word "tenths" represents that only one digit is in the decimal part.)

362.84 : three hundred sixty-two and eighty-four hundredths.

("eighty-four" is a two digit number, and the word "hundredths" represents that two digits are in the decimal part.)

17.502 : seventeen and five hundred two thousandths

("five hundred two" is a three digit number, and the word "thousandths" represents that three digits are in the decimal part.)

243.078 : Two hundred forty-three and seventy-eight thousandths

("seventy eight" is a two digit number, but the word "thousandths" signifies that three digits are in the decimal part. So, a zero should be placed before the two digit number to make three digits in the decimal part.)

.02 : two hundredths
.002 : two thousandths
.0002 : two ten thousandths

If there is no decimal part, then there is no decimal point.
20.0 = 20. = 20

Unnecessary Zeroes:
Remove zeroes from the front of whole numbers and from the back of decimal numbers.
02.010 = 2.01
20.10 = 20.1
0.20 = .2 or 0.2

(2) Rounding

When a precise calculation is not required, we round the number to estimate.

Rounding a number results in an answer approximately equal to the exact answer.

We round a number to the nearest ten, hundred, thousand, and so on, depending on the accuracy.

Rounding Method

1) If the digit to the right of the place to which we are rounding is less than 5, round down, replacing it with a 0 .

Example

① Round 23 to the nearest ten:

2 is in the tens place. Since the digit to the right of the tens place is 3 (which is less than 5), replace the 3 with a 0. The rounded result is 20. That is, $23 \approx 20$

② Round 846 to the nearest hundred:

8 is in the hundreds place. Consider the digit of 4. Since 4 is less than 5, replace 46 to 00. The rounded result is 800. That is, $846 \approx 800$

---

2) | If the digit to the right of the place to which we are rounding is 5 or more, then replace it with a 0 and add 1 to the left digit.

Example

① Round 46 to the nearest ten:

4 is in the tens place. Since the digit to the right of the tens place is the 6 (which is greater than 5), replace the 6 with a 0 and add 1 to 4. The rounded result is 50. That is, $46 \approx 50$

② Round 354 to the nearest hundred:

3 is in the hundreds place. Consider the digit of 5. Since 5 is equal to 5, replace 54 with 00 and add 1 to 3. The rounded result is 400. That is, $354 \approx 400$

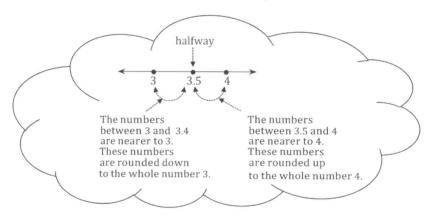

(3) Operations of Decimals

1) Adding and Subtracting Decimals

$3 = 3.0 = 3.00 = \cdots$

Basic rules

① Line up the decimal points.

② Pin the decimal point on the back of the whole number.

③ Match the number of decimal digits. That is, fill the empty decimal places with zeroes for each number to contain the same number of digits after the decimal point.

④ Adding and subtracting are exactly same as with whole numbers.

Example

$5.4 + 3.735 \Rightarrow$
$$
\begin{array}{r}
5.4 \\
+)\underline{3.735} \\
\end{array}
$$
↑
Line up
the decimal points

$\Rightarrow$
$$
\begin{array}{r}
5.400 \\
+)\underline{3.735} \\
\end{array}
$$
Fill in
the missing place values
with zeroes

$\Rightarrow$
$$
\begin{array}{r}
5.400 \\
+)\underline{3.735} \\
9.135 \\
\end{array}
$$
Add as with
whole numbers

$5.4 - 3.735 \Rightarrow$
$$
\begin{array}{r}
5.4 \\
-)\underline{3.735} \\
\end{array}
$$
↑
Line up
the decimal points

$\Rightarrow$
$$
\begin{array}{r}
5.400 \\
-)\underline{3.735} \\
\end{array}
$$
Fill in
the missing place values
with zeroes

$\Rightarrow$
$$
\begin{array}{r}
5.400 \\
-)\underline{3.735} \\
1.665 \\
\end{array}
$$
Subtract as with
whole numbers

2) Multiplying Decimals

Basic rules

① Do not line up the decimal points. Set up the problem as with whole numbers.

② To place the decimal point in the answer, count the total number of digits to the right of the decimal point in both numbers.

③ Starting from the right, count over the total number of digits and place the decimal point in the answer so that there are the same total digits after the decimal point in the answer.

Example

4.58×6.7:   4.58 has two decimal digits and 6.7 has one decimal digit.

So, the answer must have three decimal digits. $4.58 \times 6.7 = 30.686$

④ If the number of decimal digits required in the answer is higher than the total number of digits of the product, fill the empty decimal places with zeroes until the answer contains the necessary number of decimal digits.

Example

0.0008×5.2:   0.0008 has four decimal digits and 5.2 has one decimal digit.

So, the answer must have five decimal digits. But the product has only three digits.

To match the same total number of decimal places, fill two zeroes in the places to the left of the number, making it 00416.

Therefore, $0.0008 \times 5.2 = 0.00416$

⑤ To multiply a decimal by 10^n, $n = 1, 2, 3, \cdots\cdots$, move back the decimal point to the right as many as the number of zeroes in 10^n.

Example

$0.25 \times 10 = 2.5$ $(0.25 \Rightarrow 02.5 = 2.5)$
　　　　　　　　Move back 1 place

$2.5 \times 10 = 25$ $(2.5 \Rightarrow 25. = 25)$
　　　　　　　　Move back 1 place

$2.5 \times 100 = 250$ $(2.5 \Rightarrow 250. = 250)$
　　　　　　　Move back 2 places
　　　　　　　and fill zeroes

$2.5 \times 1000 = 2500$ $(2.5 \Rightarrow 2500. = 2500)$
　　　　　　　Move back 3 places
　　　　　　　and fill zeroes

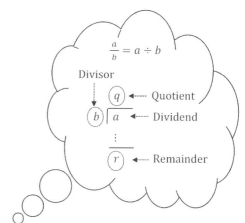

3) Dividing Decimals

Basic rules

① If a divisor contains decimal digits, transfer the divisor into a whole number by moving the decimal point to the right and then moving the decimal point of the dividend to the right by the same number of places. If possible, fill in the empty places of the dividend with zeroes to match the number of moving.

Example

$5.2 \div 2.34$

$2.34 \overline{\smash{)}5.2} \Rightarrow 234. \overline{\smash{)}5.2} \Rightarrow 234 \overline{\smash{)}520.} \Rightarrow 234 \overline{\smash{)}520}$
Move back 2 places　　Move back 2 places　　Fill in　　　　　Divide as with
to the right　　　　　to the right　　　the empty place　　a whole number
　　　　　　　　　　　　　　　　　with a zero

② Before dividing, place a decimal point in the quotient directly above the decimal point in the dividend.

2. Decimals and Fractions

(1) Terminating (Finite) Decimals

A finite decimal is a decimal which has only a finite number of values except zero after the decimal point.

For example, 0.2 , 0.23 , 0.2345

Note

① *Repeating (infinite) decimal*

A repeating decimal is a decimal which has an infinite number of values except zero after the decimal point. We place a bar over the set of numbers to be repeated indefinitely.

For example, $0.3333\cdots = 0.\overline{3}$, $2.5434343\cdots = 2.5\overline{43}$

Repeating (infinite) decimals represent rational numbers.

Rational numbers
- Integers
  - Positive integers (1, 2, 3, ⋯)
  - Zero (0)
  - Negative integers (−1, −2, − 3, ⋯)
- Non ⊤ integers (fractions)
  - Terminating decimals (finite)
  - Repeating decimals (infinite)

② *Non-repeating infinite decimals*

A non-repeating infinite decimal has an infinite number of decimal digits without any patterns.

Non-repeating infinite decimals are not considered rational numbers.

For example, $2.312347\cdots$

(2) Converting a terminating decimal into a fraction

1) All terminating decimals are converted into fractions with denominators which are powers of 10. The number of decimal digits equals to the number of the power of 10.

① If the decimal has **1** decimal digit, place the decimal digit in the numerator and place the number of 10 ($= 10^1$) in the denominator.

② If the decimal has **2** decimal digits, place the two decimal digits in the numerator and place the number of 100 ($= 10^2$) in the denominator.

Example

$$0.2 = \frac{2}{10}, \quad 0.23 = \frac{23}{100} = \frac{23}{10^2}, \quad 0.008 = \frac{8}{1000} = \frac{8}{10^3}$$

2) After converting a terminating decimal into a fraction, the reduced fraction to lowest term has a denominator whose prime factors are only 2 or 5.

Example

$$0.2 = \frac{2}{10} = \frac{2}{2\times5} = \frac{1}{5} \qquad 0.23 = \frac{23}{100} = \frac{23}{10^2} = \frac{23}{(2\times5)^2} = \frac{23}{2^2\times5^2}$$

$$0.008 = \frac{8}{1000} = \frac{8}{10^3} = \frac{8}{(2\times5)^3} = \frac{8}{2^3\times5^3} = \frac{1}{5^3}$$

3) If the decimal is a mixed number, the decimal part is separated from its whole number part. After converting the decimal part into a fraction, the whole part of the decimal remains the whole number part of the equivalent mixed number.

Example

$2.5 = 2 + 0.5$ and $0.5 = \dfrac{5}{10} = \dfrac{1}{2}$ $\therefore 2.5 = 2\dfrac{1}{2}$

$4.008 = 4 + 0.008$ and $0.008 = \dfrac{8}{10^3} = \dfrac{8}{(2\times5)^3} = \dfrac{2^3}{2^3\times5^3} = \dfrac{1}{5^3}$ $\therefore 4.008 = 4\dfrac{1}{5^3}$

(3) Converting a repeating decimal into a fraction

Converting method

1) $\boxed{0.\overline{a} = \dfrac{a}{9}}$

Let $x = 0.\overline{a} = 0.aaa\cdots$ Then, $10x = a.aaa\cdots$

So, $10x = a.aaa\cdots$

$\underline{-)\ \ x = 0.aaa\cdots}$

$9x = a$; $x = \dfrac{a}{9}$ $\therefore 0.\overline{a} = \dfrac{a}{9}$

2) $\boxed{0.a\overline{b} = \dfrac{ab-a}{90}}$

Let $x = 0.a\overline{b} = 0.abbb\cdots$ Then, $10x = a.bbb\cdots$ and $100x = ab.bbb\cdots$

So, $100x = ab.bbb\cdots$

$\underline{-)\ 10\,x = a.bbb\cdots}$

$90x = ab - a$; $x = \dfrac{ab-a}{90}$ $\therefore 0.a\overline{b} = \dfrac{ab-a}{90}$

Similarly, $\boxed{0.a\overline{bc} = \dfrac{abc-a}{990}}$ and $\boxed{0.a\overline{bcd} = \dfrac{abcd-ab}{9900}}$

3) For the denominator, place the number 9 as many times as the number of repeating digits (the number of digits under the " $\overline{}$ ") in the equivalent fraction . Then, place the number 0 as many times as the number of non-repeating decimal digits (the number of decimal digits which are not under the " $\overline{}$ ") right after the number 9.

For the numerator, subtract the non-repeating digits from the whole digits, ignoring the decimal point.

Example $23.4\overline{56}$

For the denominator, count how many number of digits are repeating (how many digits are under the " $\overline{}$ "). Since two digits, 5 and 6, are repeating and one digit, 4, is a non-repeating decimal digit, place 9 two times and 0 one time after the 9 in the denominator.

For the numerator, subtract the non-repeating digits, 234 from the whole digits, 23456.

Then, $23.4\overline{56} = \dfrac{23456 - 234}{990} = \dfrac{23222}{990}$

(4) Converting a fraction into a decimal

Basic rules

① Reduce the fraction to lowest term.

② Check whether the denominator has only the prime factors 2 or 5.

If it does, the fraction can be converted into a terminating decimal.

If the denominator has factors other than 2 or 5, the fraction cannot be written as a terminating decimal.

$2 \times 5 = 10$
$2^2 \times 5^2 = (2 \times 5)^2 = 10^2$
$2^3 \times 5^3 = (2 \times 5)^3 = 10^3$

Example Express $\dfrac{3}{50}$ as a decimal.

First, find the factors of the denominator using prime factorization.

Note that: $\dfrac{3}{50} = \dfrac{3}{2 \times 5^2}$.

To have equal powers of the prime factors 2 and 5, multiply both the numerator and denominator by 2. Then the denominator can be written as a power of $10 (= 2 \times 5)$.

That is, $\dfrac{3}{50} = \dfrac{3}{2 \times 5^2} = \dfrac{3 \times 2}{2 \times 5^2 \times 2} = \dfrac{6}{2^2 \times 5^2} = \dfrac{6}{(2 \times 5)^2} = \dfrac{6}{10^2} = \dfrac{6}{100} = 0.06$

Therefore, $\dfrac{3}{50} = 0.06$

$\dfrac{6}{10} = \dfrac{6}{10^1} = 0.6$ (**1** decimal digit)

$\dfrac{6}{100} = \dfrac{6}{10^2} = 0.06$ (**2** decimal digits)

$\dfrac{6}{1000} = \dfrac{6}{10^3} = 0.006$ (**3** decimal digits)

Example Express $\dfrac{7}{125}$ as a decimal.

$\dfrac{7}{125} = \dfrac{7}{5^3} = \dfrac{7 \times 2^3}{5^3 \times 2^3} = \dfrac{7 \times 2^3}{(5 \times 2)^3} = \dfrac{56}{10^3} = \dfrac{56}{1000} = 0.056$

Example Express $\dfrac{1}{12}$ as a decimal.

Since $12 = 4 \times 3 = 2^2 \times 3$, the denominator of the fraction has a prime factor 3, which is a number other than 2 or 5. So the fraction cannot be written as a terminating decimal.

In fact, $1 \div 12 = 0.08333\cdots = 0.08\overline{3}$ (Repeating infinite decimal).

6-2 Other Algebraic Expressions

1. Ratio and Proportion

> A rate is a fixed ratio between two things.

(1) Ratio

The *ratio* describes a relationship between two numbers. Ratios are written as fractions in lowest terms. Unlike fractions, we don't convert the improper fractions into mixed numbers. We write the terms of the ratio in three different ways :

$$1 \text{ to } 2, \quad 1:2, \quad \text{or} \quad \frac{1}{2} \text{ (means 1 out of 2)}$$

>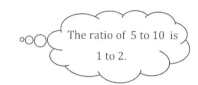
> The ratio of 5 to 10 is 1 to 2.

(2) Proportion

Proportion is an equality of a relationship between two sets of numbers.

In an equation, $\frac{a}{b} = \frac{c}{d}$ means a out of b is proportional (or equal) to c out of d.

> A proportion contains two ratios that are equal.

Note:

The rational numbers $\frac{a}{b}$ and $\frac{c}{d}$ are equal. $\xLeftrightarrow[\text{if and only if}]{}$ $ad = bc$

To solve proportions, the cross product is used.

> For any numbers a and b, $(a \neq 0, b \neq 0)$
> $\frac{a}{b} = \frac{c}{d} \Rightarrow ad = bc$
> In a proportion, the cross products are equal.

Example

Solve the equation: $\frac{2}{3} = \frac{x}{24}$

$\frac{2}{3} = \frac{x}{24} \quad \Rightarrow \quad 2 \cdot 24 = 3 \cdot x$ (by cross product) $\Rightarrow \quad 48 = 3x$

To isolate the variable on the side of the equal sign, divide both sides of the equation by the coefficient of x. Then, $\frac{48}{3} = \frac{3x}{3} \quad \therefore 3x = 48 \; ; \; x = 16$

Or, we can get it easily by ratio:

$$\frac{2}{3} = \frac{x}{24} \quad \therefore x = 16$$
$\times 8$

> An equation is an expression using an equal sign between two expressions. A coefficient is a number that is multiplied by a variable.

Example

Solve the equation: $\frac{x+5}{2} = 3$

To use the cross product in the proportion, the whole number 3 must be converted into a fraction : $\frac{3}{1}$. Then, $\frac{x+5}{2} = \frac{3}{1} \; ; \; (x + 5) \cdot 1 = 2 \cdot 3 \; ; \; x + 5 = 6$

To isolate the variable on one side of the equal sign, subtract both sides of the equation by the constant 5. Then, $x + 5 - 5 = 6 - 5 \quad \therefore x = 1$

2. Percent

A *percent* is a ratio of a number to 100 using the symbol %.

For example, $25\% = \dfrac{25}{100}$; 25 out of 100

A percentage is an amount stated as if it is a part of a whole, which is 100.

$25\% = \dfrac{25}{100} = \dfrac{1}{4}$ (reduce to the lowest term)

(1) Conversion

1) Converting a Percentage into a Fraction

Removing the symbol %, divide the number by 100 or multiply the number by $\dfrac{1}{100}$ and reduce to the lowest term.

Example

$$25\% = \dfrac{25}{100} = \dfrac{1}{4}, \qquad \dfrac{1}{4}\% = \dfrac{\frac{1}{4}}{100} = \dfrac{1}{400}$$

$$6\dfrac{2}{3}\% = \dfrac{20}{3}\% = \dfrac{20}{3} \times \dfrac{1}{100} = \dfrac{1}{15}, \qquad 25\% \text{ of } 200 = \dfrac{25}{100} \times 200 = 50$$

2) Converting a Fraction into a Percentage

For example, to express $\dfrac{1}{5}$ as a percent, set up the equation: $\dfrac{1}{5} = x\% = \dfrac{x}{100}$

By cross product, $5x = 100 \quad \therefore \ x = \dfrac{100}{5} = 20 \qquad$ Therefore, $\dfrac{1}{5} = 20\%$

3) Converting a Percentage into a Decimal

Removing the symbol "%", move the decimal point two places to the left of the number.

If there are empty places after the moved decimal point, fill them with zeroes.

Example

$0.5\% \xrightarrow[\text{Remove \%}]{} 0.5 \implies .\underset{\substack{\text{Move 2 places} \\ \text{to the left}}}{\curvearrowleft\curvearrowleft}0.5 \xrightarrow[\text{Fill with a zero}]{} .005 \quad \therefore 0.5\% = .005$

$5\% \xrightarrow[\text{Remove \%}]{} 5. \implies .\underset{\substack{\text{Move 2 places} \\ \text{to the left}}}{\curvearrowleft\curvearrowleft}5. \xrightarrow[\text{Fill with a zero}]{} .05 \quad \therefore 5\% = .05$

$50\% \xrightarrow[\text{Remove \%}]{} 50. \implies .\underset{\substack{\text{Move 2 places} \\ \text{to the left}}}{\curvearrowleft\curvearrowleft}50. \implies .50(= .5) \quad \therefore 50\% = .5$

> Zeroes from the front of the whole number and from the back of the decimal number can be removed.
>
> **020.050** = 20.05

4) Converting a Decimal into a Percentage

Move the decimal point two places to the right of the number. If there are empty places before the moved decimal point, fill them with zeroes. Add the symbol "%".

Example

$5 = 5.$ $\implies$ $5.$ $\implies$ $5\ 0\ 0.$ $\quad \therefore 5 = 500\%$

Move 2 places   Fill empty places
to the right   with zeroes

2.5 $\implies$ $2.5.$ $\implies$ $250.$ $\quad \therefore 2.5 = 250\%$

Move 2 places   Fill empty place
to the right   with a zero

0.25 $\implies$ $0.25.$ $\implies$ $025.$ $\quad \therefore 0.25 = 25\%$

Move 2 places
to the right

0.025 $\implies$ 0.025 $\implies$ 002.5 $\quad \therefore 0.025 = 2.5\%$

Move 2 places
to the right

(2) Interest

Interest is the percent of a principal over a period of time.

1) Simple Interest

The principal is an amount of money lent, put into a bank, etc., on which interest is paid.

Interest is money that is added to the principal.

Simple interest is the percent of the principal that has to be paid by the consumer or by the bank.

Simple interest is calculated only on the principal. So, it can be written as:

$$I = prt$$

where I is interest, p is the principal, r is the annual interest rate expressed as a decimal, and t is the period of time in years (number of years).

Example

What is the simple interest on a $500 loan at 8.5% interest for $\frac{1}{2}$ year?

How much money will be paid on the interest?

$I = prt\ = 500 \times 0.085 \times 0.5 = 21.25$

Since $500 + 21.25 = 521.25$, $\ 521.25 will be paid.

2) Compound Interest

Compound interest is calculated on the principal and the previously earned interest.

Thus, it can be written as

$$A = p\left(1 + \frac{r}{n}\right)^{nt}$$

where A is the balance (total amount = principal + interest), p is the principal (starting balance), r is the annual interest rate expressed as a decimal, n is the number of times the interest is compounded annually, and t is the period of time in years.

Example

You deposit $700 in a saving account at 6% annual interest rate, compounded monthly. What will be the balance of the account 6 months later? (Round the answer to the hundredths place.)

Since $p = 700$, $r = 6\% = 0.06$, $n = 12$, and $t = 6$ months $= \frac{6}{12}$,

$$A = p\left(1 + \frac{r}{n}\right)^{nt} = 700\left(1 + \frac{0.06}{12}\right)^{12 \cdot \frac{6}{12}} = 700\,(1 + 0.005\,)^6 = 721.2642566 \approx 721.26$$

Therefore, the balance of the account will be $721.26

3. Scientific Notations

To express very large and very small numbers, you use scientific notations:

(1) Scientific Notations

1) For large numbers

Move the decimal point to the left until only one non-zero number remains left of the decimal point. Multiply the resulting decimal by 10^n, where n is the number of digits you moved the decimal to the left.

Note: $10 = 10^1$, $100 = 10^2$, $1000 = 10^3$, $10000 = 10^4$, $\cdots\cdots$

Example $230000 = 230000. = 2\,3,0\,0\,0\,0. = 2.3 \times 10^5$

Only one
non−zero number

Move the decimal point
5 places to the left

If you move the decimal point to the left
$\Rightarrow$ positive power of 10

$$-4{,}500{,}000 = -4{,}500{,}000. = -4.5 \times 10^6$$

Unnecessary zeroes:

$2 = 2.0 = 2.00 = \cdots = 2.$

2) For small numbers

Move the decimal point to the right until only one non-zero number appears left of the decimal point. Multiply the resulting decimal by 10^{-n}, where n is the number of digits you moved the decimal to the right.

Note: $0.1 = 10^{-1}$, $0.01 = 10^{-2}$, $0.001 = 10^{-3}$, $0.0001 = 10^{-4}$, ······

Example $0.000023 = 0.0\,0\,0\,0\,2\,3 = 2.3 \times 10^{-5}$

Move the decimal point 5 places to the right

Only one non−zero number

> If you move the decimal point to the right ⇒ negative power of 10

$$-0.0000045 = -4.5 \times 10^{-6}$$

> Unnecessary zeroes: $2 = 02 = 0002 = \cdots$

(2) Operations in Scientific Notations

1) Adding and subtracting numbers in scientific notation:

To add or subtract numbers in scientific notation, their powers of 10 must be the same. Use the distributive property to simplify the expressions. If the result is not in scientific notation, adjust it.

> Distributive property : $a \cdot c + b \cdot c = (a + b) \cdot c$

Example

$$7.4 \times 10^2 + 4.8 \times 10^2 = (7.4 + 4.8) \times 10^2 = 12.2 \times 10^2 = 1.22 \times 10^1 \times 10^2 = 1.22 \times 10^3$$

Two non-zero numbers   One non-zero number   Scientific notation

$$57.4 \times 10^2 - 4.8 \times 10^3 = 5.74 \times 10 \times 10^2 - 4.8 \times 10^3 = 5.74 \times 10^3 - 4.8 \times 10^3$$
$$= (5.74 - 4.8) \times 10^3 = 0.94 \times 10^3 = 9.4 \times 10^{-1} \times 10^3 = 9.4 \times 10^2$$

Two non-zero numbers   One non-zero number   Scientific notation

> Only one non-zero digit remains left of the decimal point.

2) Multiplying and dividing numbers in scientific notation

To multiply or divide numbers in scientific notation, their powers of 10 don't need to be the same. Separate the expressions into decimal parts and exponents. Then, multiply or divide the decimal parts and multiply or divide the exponents, separately. Finally, multiply both results. If the result is not in scientific notation, adjust it.

Example

$$(2.5 \times 10^2) \cdot (6.3 \times 10^{-5}) = (2.5 \times 6.3) \times (10^2 \times 10^{-5})$$

$$= 15.75 \times 10^{-3} = (1.575 \times 10^1) \times 10^{-3} = 1.575 \times 10^{-2}$$

Two non-zero numbers   One non-zero number   Scientific notation

Same base property:

$$a^m \cdot a^n = a^{m+n}$$

$$\frac{a^m}{a^n} = a^{m-n}$$

$$(2.184 \times 10^{-12}) \div (5.2 \times 10^{-7}) = (2.184 \div 5.2) \times (10^{-12} \div 10^{-7})$$

$$= 0.42 \times 10^{-5} = 4.2 \times 10^{-1} \times 10^{-5} = 4.2 \times 10^{-6}$$

Two non-zero numbers   One non-zero number   Scientific notation

$$ab \div cd = \frac{ab}{cd} = \frac{a}{c} \times \frac{b}{d}$$

Exercises

#1 Name the place of the underlined digit.

(1) $\underline{2}$34.2

(4) 35.134$\underline{8}$

(7) 6$\underline{2}$5.34

(2) 15$\underline{6}$.9

(5) 2.53$\underline{2}$

(3) 23. $\underline{5}$9

(6) 4.5$\underline{2}$1

#2 Round each decimal to the nearest whole number, tenths, hundredths, and thousandths separately.

(1) 44.5362

(3) 2.0534

(5) 19.9995

(2) 32.4997

(4) 1.2209

#3 Order the following numbers from least to greatest:

(1) 0.5 , -0.24 , 0.48 , -0.024 , 0.418 , 0.05

(2) -0.3 , -0.03 , 0.31 , 0.13 , 0.013 , -0.13

(3) 2.4 , 2.04 , 2.41 , 2.39 , -0.24 , -0.21

(4) 0.05 , 0.49 , 0.409 , 0.41 , 0.419 , 0.5

(5) -0.6 , -0.06 , -0.61 , -0.59 , -0.061 , -0.509

#4 Calculate the sum or difference for the following expressions:

(1) $2.3 + 5.84$

(8) $0.2 - 5.94$

(15) $12.9+(-15)$

(2) $3.45 + 2.9$

(9) $4.538 - 35.6$

(16) $25.8-(-5.29)$

(3) $0.2 + 5.94$

(10) $5.7 - 0.49$

(17) $1+(-0.99)$

(4) $4.538 + 35.6$

(11) $-5.8 - 8.5$

(18) $6.5+(-6.5)$

(5) $5.7 + 0.49$

(12) $-3.9 + 6.4$

(19) $-100 - (-0.11)$

(6) $2.3 - 5.84$

(13) $-2.5 - (+3.6)$

(20) $-2.5 - (+3.72)$

(7) $3.45 - 2.9$

(14) $-0.002 - (-3.24)$

#5   Calculate the product or division for the following expressions:

(1) 0.2×0.5

(2) 0.25×0.4

(3) 2.5×3.4

(4) 0.33×0.03

(5) 3.4×12

(6) 2×0.005

(7) -2.6×0.8

(8) -5.7×-2.3

(9) -0.1×-0.01

(10) 1.25×100

(11) 0.004×100

(12) 0.3×10^3

(13) 0.0067×10^2

(14) $3.42 \div 2$

(15) $20.8 \div 0.2$

(16) $8.4 \div 0.04$

(17) $-12 \div 0.05$

(18) $16 \div (-0.5)$

(19) $-0.025 \div (-0.02)$

(20) $3 \div 0.03$

#6   Simplify each expression and round to the nearest hundredths.

(1) $3 \times 2.5 \div 7$

(2) $5 - 0.25 \div 3$

(3) $-6 \times 0.03 \div 8$

(4) $4 \div (-5.3) \times 0.2$

(5) $-2 - (-0.002) \div (-0.5)$

#7   Convert the following decimals into fractions and reduce the fraction to lowest terms:

(1) 0.5

(2) 3.45

(3) 0.032

(4) 2.05

(5) 10.46

(6) 5.025

(7) 4.56

(8) 6.55

(9) 7.008

(10) 9.16

#8   Convert each repeating decimal into a fraction.

(1) $0.\overline{2}$

(2) $0.\overline{02}$

(3) $0.3\overline{4}$

(4) $0.5\overline{67}$

(5) $0.48\overline{6}$

(6) $2.\overline{3}$

(7) $3.4\overline{2}$

(8) $8.5\overline{67}$

(9) $3.4\overline{56}$

(10) $5.\overline{123}$

#9 Convert each fraction into a terminating decimal.

(1) $\dfrac{3}{8}$ (5) $\dfrac{5}{12}$ (9) $\dfrac{36}{80}$

(2) $\dfrac{4}{25}$ (6) $\dfrac{34}{60}$ (10) $\dfrac{1}{250}$

(3) $\dfrac{6}{20}$ (7) $\dfrac{24}{50}$

(4) $\dfrac{3}{40}$ (8) $\dfrac{7}{200}$

#10 Express each ratio as a fraction in lowest term.

(1) 3 to 12   (5) $\dfrac{1}{2}$ to 0.2   (8) 2.4 to 0.02

(2) 9 to 3   (9) 0.8 to 2

(3) 15 : 24   (6) $3\dfrac{1}{2} : 5.2$ (10) $0.03 : \dfrac{3}{5}$

(4) 32 to 14   (7) $\dfrac{3}{4} : 1\dfrac{2}{3}$

#11 Solve each problem.

(1) 3 pencils cost 45 cents. How much do 5 pencils cost?

(2) There are 5 nickels for every 24 dimes. How many nickels will there be for 96 dimes?

(3) A teacher supplies 6 books for each 20 students. How many students are needed for 27 books to be supplied?

(4) 8 apples cost 2 dollars. How much do 10 apples cost?

(5) There are 27 cookies in 3 bowls. How many cookies would there be in 7 bowls?

#12 Solve the following proportion for x:

(1) $\dfrac{5}{3} = \dfrac{x}{6}$ (5) $\dfrac{3}{2.5} = \dfrac{x}{5}$ (9) $\dfrac{6}{x} = \dfrac{3}{7}$

(2) $\dfrac{x}{4} = \dfrac{6}{5}$ (6) $\dfrac{0.2}{3.2} = \dfrac{x}{4}$ (10) $\dfrac{x}{0.04} = \dfrac{5}{4}$

(3) $\dfrac{3}{x} = \dfrac{24}{8}$ (7) $\dfrac{x}{6} = \dfrac{27}{54}$

(4) $\dfrac{2}{7} = \dfrac{16}{x}$ (8) $\dfrac{0.05}{0.2} = \dfrac{x}{0.8}$

#13   Convert the following percentages into fractions and reduce the fractions to lowest terms:

(1) 30%

(2) 120%

(3) $\frac{1}{2}$%

(4) $25\frac{1}{3}$%

(5) 3.5%

(6) 75%

#14   Solve each expression.

(1) 65 % of 140

(2) 25 is 40 % of what number?

(3) 70 % of what number is 28?

(4) What percent is 20 of 25?

(5) What number is $2\frac{1}{3}$ % of 240?

(6) What percent is 5 of 200?

(7) What percent of 20 is 10?

#15   State the change (increase or decrease) of each percent.

(1) What is the percent change from 12 to 3?

(2) What is the percent change from 3 to 12?

#16   Convert each fraction into a percentage.

(1) $\frac{1}{4}$

(2) $\frac{3}{10}$

(3) $\frac{1}{25}$

(4) $\frac{3}{50}$

(5) $\frac{12}{5}$

(6) $\frac{1}{100}$

(7) $\frac{25}{4}$

(8) 1

(9) $\frac{3}{200}$

(10) $\frac{9}{15}$

#17   Convert each percentage into a decimal.

(1) 25 %

(2) 2.5 %

(3) 0.25 %

(4) 250 %

(5) 22.5 %

(6) 1 %

(7) 10 %

(8) 100 %

(9) 50 %

(10) 0.5 %

#18 Convert each decimal into a percentage.

(1) 0.1

(2) 12.5

(3) 0.84

(4) 45.2

(5) 0.02

(6) 0.001

(7) 0.478

(8) 1

(9) 0.01

(10) 0.5

#19 Given the principal, p, the rate of interest, r, and time, t, find the simple interest.

(1) $p = \$500$, $r = 10\%$, $t = 2$ years

(2) $p = \$1000$, $r = 2.5\%$, $t = 1\frac{1}{2}$ years

(3) $p = \$600$, $r = 3\frac{3}{4}\%$, $t = 2\frac{1}{4}$ years

#20 Find the principal for the following (round the answer to the whole number):

(1) When $50 of interest is paid in 3 months at a rate of 8 %

(2) When $60 of interest is paid in $2\frac{1}{2}$ years at a rate of 3.5 %

(3) When $10 of interest is paid in 8 months at a rate of $4\frac{1}{2}$ %

#21 Find the total amount, rounding the answer to the hundredths.

(1) Principal = $ 500   Rate = 30% compounded annually   Time = 3 years

(2) Principal = $ 500   Rate = 5% compounded monthly   Time = 6 months

(3) Principal = $ 300   Rate = 6% compounded semi-annually   Time = 3 years

(4) Principal = $ 300   Rate = 4% compounded quarterly   Time = 2 years

(5) Principal = $ 1000   Rate = $6\frac{3}{4}$ % compounded monthly   Time = $\frac{1}{4}$ of a year

#22 Express the following numbers using scientific notation:

(1) 125

(2) 125000

(3) $125 \times 2 \times 10^5$

(4) -2400

(5) 0.00125

(6) 0.00000125

(7) $0.125 \times 2 \times 10^5$

(8) -0.0024

#23 Simplify the following expressions using scientific notation:

(1) $2.54 \times 10^2 + 3.2 \times 10^3$

(2) $5.21 \times 10^{-3} - 4.8 \times 10^{-3}$

(3) $(3.2 \times 10^{-5}) \times (6.4 \times 10^8)$

(4) $(1.12 \times 10^3) \div (3.2 \times 10^{-6})$

Chapter 7. Monomials and Polynomials

7-1 Exponents

1. Definition

(1) $a \cdot a \cdot a \cdots a = a^n$: *Exponent form*

(2) *Base*: The number or letter to multiply by itself

(3) *Exponent (Power)*: The number of times to multiply the base by itself

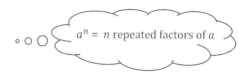

$a^n = n$ repeated factors of a

Example

$3 \cdot 3 \cdot 3 \cdot 3 = 3^4$: 3 is the base and 4 is the exponent (or power)

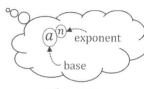

a^n — exponent; base

2. Rules of Exponents

For any real number $a(\neq 0)$ and positive integers m and n,

$1 = 1^1$

Any number to the first power is equal to itself.

$2 = 2^1$; $3 = 3^1$; $\cdots$

> **(1) Addition of Exponents**
>
> $$a^m \cdot a^n = \underbrace{a \cdot a \cdots a \cdot a}_{m \text{ times}} \cdot \underbrace{a \cdot a \cdots a \cdot a}_{n \text{ times}} = a^{m+n}$$
>
> **(2) Multiplication of Exponents**
>
> $$(a^m)^n = \underbrace{a^m \cdot a^m \cdots a^m \cdot a^m}_{n \text{ times}} = a^{\overbrace{m+m+\cdots+m}^{n \text{ times}}} = a^{mn}$$
>
> **(3) Division of Exponents**
>
> $$a^m \div a^n = \frac{a^m}{a^n} = \frac{\overset{m}{\overbrace{a \cdots a}}}{\underset{n}{\underbrace{a \cdots a}}} = \begin{cases} a^{m-n}, & m > n \\ 1, & m = n \\ \frac{1}{a^{n-m}}, & m < n \end{cases}$$

$a^4 \div a^2 \neq a^{4 \div 2}$

Example

$$a^3 \div a^2 = \frac{a^3}{a^2} = \frac{a \cdot a \cdot a}{a \cdot a} = a = a^{3-2}; \quad a^3 \div a^3 = \frac{a^3}{a^3} = \frac{a \cdot a \cdot a}{a \cdot a \cdot a} = 1; \quad a^2 \div a^3 = \frac{a^2}{a^3} = \frac{a \cdot a}{a \cdot a \cdot a} = \frac{1}{a} = \frac{1}{a^{3-2}}$$

Note: When you move the exponential terms across the fraction bar, move the term with the smaller exponent.

For example,

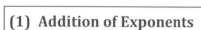

$$a^3 \div a^2 = \frac{a^3}{a^2} = \frac{a^3 \cdot a^{-2}}{1} = a^1 = a; \qquad a^2 \div a^3 = \frac{a^2}{a^3} = \frac{1}{a^3 \cdot a^{-2}} = \frac{1}{a^1} = \frac{1}{a}$$

$$a^{-2} \div a^{-3} = \frac{a^{-2}}{a^{-3}} = \frac{a^{-2} \cdot a^3}{1} = \frac{a^1}{1} = a^1 = a; \qquad a^{-3} \div a^{-2} = \frac{a^{-3}}{a^{-2}} = \frac{1}{a^{-2} \cdot a^3} = \frac{1}{a^1} = \frac{1}{a}$$

3. Distributive Properties of Exponents

(1)
$$(ab)^m = \underbrace{(ab) \cdot (ab) \cdots (ab) \cdot (ab)}_{m \text{ times}}$$
$$= \underbrace{(a \cdot b) \cdot (a \cdot b) \cdots (a \cdot b) \cdot (a \cdot b)}_{m \text{ times}} = \underbrace{a \cdot a \cdots a \cdot a}_{m \text{ times}} \cdot \underbrace{b \cdot b \cdots b \cdot b}_{m \text{ times}} = a^m \cdot b^m$$

(2)
$$\left(\frac{a}{b}\right)^m = \underbrace{\left(\frac{a}{b}\right) \cdot \left(\frac{a}{b}\right) \cdots \left(\frac{a}{b}\right) \cdot \left(\frac{a}{b}\right)}_{m \text{ times}} = \frac{\overbrace{a \cdots a}^{m}}{\underbrace{b \cdots b}_{m}} = \frac{a^m}{b^m}$$

Example

$$\left(\frac{a}{b}\right)^2 = \left(\frac{a}{b}\right) \cdot \left(\frac{a}{b}\right) = \frac{a \cdot a}{b \cdot b} = \frac{a^2}{b^2}$$

Note

- $\left(-3\frac{a}{b}\right)^2 = (-3)^2 \cdot \left(\frac{a}{b}\right)^2 = 9\frac{a^2}{b^2}$

- $(a^l b^m)^n = a^{ln} b^{mn}$, positive integers l, m, n, $a \neq 0$, $b \neq 0$

- $\left(\frac{a^l}{b^m}\right)^n = \frac{a^{ln}}{b^{mn}}$

4. Expanding Exponents

(1) $a^0 = 1$

(2) $a^{-m} = \frac{1}{a^m}$

(3) $a^{m+1} - a^m = a^m(a - 1)$

$$1 = a^2 \div a^2 = a^{2-2} = a^0$$
$$1^0 = 1 \; ; \; 2^0 = 1; \; 3^0 = 1$$
But $0^0 \neq 1$ ($\because 0^0$ is undefined.)

$$a^{m+1} = a^m \cdot a^1 = a^m \cdot a$$

Negative exponents

1) $a^{-m} = \frac{1}{a^m}$

Example

$\frac{a^2}{a^5} = a^{2-5} = a^{-3}$, by the same base quotient rule, and

$$\frac{a^2}{a^5} = \frac{a \cdot a}{a \cdot a \cdot a \cdot a \cdot a} = \frac{1}{a \cdot a \cdot a} = \frac{1}{a^3}$$

$$\therefore \quad a^{-3} = \frac{1}{a^3}$$

2) $\boxed{\dfrac{1}{a^{-m}} = a^m}$

$\because$ Since $a^{-m} = \dfrac{1}{a^m}$,

$\dfrac{1}{a^{-m}} = \dfrac{1}{\frac{1}{a^m}}$, by substituting $\dfrac{1}{a^m}$ for a^{-m}

$= \dfrac{\frac{1}{1}}{\frac{1}{a^m}} = \dfrac{1 \cdot a^m}{1 \cdot 1} = \dfrac{a^m}{1} = a^m$

$\dfrac{\frac{a}{b}}{\frac{c}{d}} = \dfrac{a \cdot d}{b \cdot c} = \dfrac{ad}{bc}$ or $\dfrac{\frac{a}{b}}{\frac{c}{d}} = \dfrac{a}{b} \cdot \dfrac{d}{c} = \dfrac{ad}{bc}$, multiply by flipping

3) $\boxed{\dfrac{1}{a^x} = a^{-x} \; ; \quad \dfrac{1}{a^{-x}} = a^x \; ; \quad \dfrac{1}{a^{-m}} = a^m \;\; \text{(opposite power)}}$

1 over any term raised to a power(exponent) is the same as that term raised to the opposite power (exponent).

4) $\boxed{a^m \cdot a^{-m} = a^{m-m} = a^0 = 1 \;\; \text{or} \;\; a^m \cdot a^{-m} = a^m \cdot \dfrac{1}{a^m} = 1}$

a^{-m} is the multiplicative inverse (or reciprocal) of a^m.

7-2 Multiplying and Dividing Monomials

A monomial is an expression consisting of only one term which is product of numbers and variables.

Solving steps

Step 1. Solve parentheses using the rules of exponents

Step 2. $\div a \; \longrightarrow \; \times \dfrac{1}{a}$ (Reciprocal)

Step 3. Determine the sign: Number of $(-)$ sign is even $\Rightarrow (+)$

Number of $(-)$ sign is odd $\Rightarrow (-)$

Step 4. Calculate the exponents

$3x^2$: Monomial of degree 2 in x

$2x$: Monomial of degree 1 in x

Example

1. $-2a \cdot 3ab = (-2 \cdot 3) \cdot (a \cdot ab) = -6 \cdot a^2 b = -6\,a^2 b$

2. $-2a \div 3ab = -2a \cdot \dfrac{1}{3ab} = \dfrac{-2a}{3ab} = \dfrac{-2}{3} \cdot \dfrac{a}{ab} = -\dfrac{2}{3} \cdot \dfrac{1}{b} = -\dfrac{2}{3b}$

3. $3a \cdot a^2 \div (3a^2)^3 = 3a \cdot a^2 \div 3^3 a^6 = 3a \cdot a^2 \cdot \dfrac{1}{3^3 a^6} = \dfrac{3}{3^3} \cdot \dfrac{a^3}{a^6} = \dfrac{1}{3^2} \cdot \dfrac{1}{a^3} = \dfrac{1}{9a^3}$

$\boxed{\begin{array}{l} a \div b \div c = a \cdot \dfrac{1}{b} \cdot \dfrac{1}{c} = \dfrac{a}{bc}, \, (b \neq 0), (c \neq 0) \\[2mm] a \div \dfrac{b}{c} = a \cdot \dfrac{c}{b} = \dfrac{ac}{b}, \, (b \neq 0), (c \neq 0) \end{array}}$

7-3 Polynomials

A polynomial is an expression of two or more terms combined by addition and/or subtraction.

1. Adding and Subtracting Polynomials

Steps to Solve

Step 1. Remove the parentheses ( ) or other enclosure marks (brace{ }, bracket[ ], absolute value sign | |, etc.) in order from the innermost enclosure marks to the outermost ones.

Step 2. Regroup the like terms.

Step 3. Simplify by combining the expressions.

Note : (1) *A polynomial is an expression which is the sum of monomials.*

For example, $3x^2 + 2x$, $2x - 1$

(2) *The terms in a polynomial are ordered from the greatest exponent to the least (decreasing degree in variable).*

$$a_n x^n + a_{n-1} x^{n-1} + \cdots\cdots + a_2 x^2 + a_1 x^1 + a_0$$

For example, $x^2 + 2x^3 + x + 4 \rightarrow 2x^3 + x^2 + x + 4$

x^1 $4x^0$

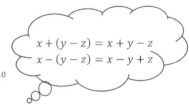

$x + (y - z) = x + y - z$
$x - (y - z) = x - y + z$

Example

1. $(2x + 3y) - (x + 2y) = 2x + 3y - x - 2y = (2x - x) + (3y - 2y) = x + y$

2. $\dfrac{x+2y}{2} - \dfrac{2x-y}{3} = \dfrac{3(x+2y)}{6} - \dfrac{2(2x-y)}{6} = \dfrac{3x+6y-4x+2y}{6} = \dfrac{(3x-4x)+(6y+2y)}{6} = \dfrac{-x+8y}{6}$

3. $2x - [3y - \{x - (2x - 3y) + 2y\} - 5] = 2x - [3y - \{x - 2x + 3y + 2y\} - 5]$

$$= 2x - [3y - x + 2x - 3y - 2y - 5]$$

$$= 2x - 3y + x - 2x + 3y + 2y + 5$$

$$= (2x + x - 2x) + (-3y + 3y + 2y) + 5$$

$$= x + 2y + 5$$

4. $2(x^2 - 3x + 5) - 3(2x^2 - x + 3) = 2x^2 - 6x + 10 - 6x^2 + 3x - 9$

$$= (2x^2 - 6x^2) + (-6x + 3x) + (10 - 9)$$

$$= -4x^2 - 3x + 1$$

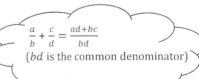

$\dfrac{a}{b} + \dfrac{c}{d} = \dfrac{ad+bc}{bd}$
(bd is the common denominator)

Note: *Linear form : when the degree of* $ax + b$ *is* 1.

Quadratic form : when the degree of $ax^2 + bx + c$ *is* 2.

Cubic form : when the degree of $ax^3 + bx^2 + cx + d$ *is* 3.

For addition,

Commutative property: $a + b = b + a$

Associative property: $(a + b) + c = a + (b + c)$

2. Multiplying and Dividing a Polynomial by a Monomial

To multiply or divide a polynomial by a monomial, use the distributive property and the rules of exponents.

$(1)\ a \cdot (b + c) = a \cdot b + a \cdot c \qquad (a + b) \cdot c = a \cdot c + b \cdot c$

$(2)\ (a + b) \div c = (a + b) \cdot \dfrac{1}{c} = \left(a \cdot \dfrac{1}{c}\right) + \left(b \cdot \dfrac{1}{c}\right) = \dfrac{a}{c} + \dfrac{b}{c}$

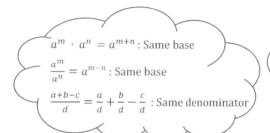

$a^m \cdot a^n = a^{m+n}$: Same base

$\dfrac{a^m}{a^n} = a^{m-n}$: Same base

$\dfrac{a+b-c}{d} = \dfrac{a}{d} + \dfrac{b}{d} - \dfrac{c}{d}$: Same denominator

For Multiplication,

Commutative property : $a \cdot b = b \cdot a$

Associative property : $a \cdot (b \cdot c) = (a \cdot b) \cdot c$

Distributive property : $a \cdot (b + c) = a \cdot b + a \cdot c$

$(b + c) \cdot a = b \cdot a + c \cdot a$

Steps to Solve

Step 1. Remove the parentheses ( ) or other enclosure marks (brace{ }, bracket[ ], absolute value sign | |, etc.) in order from the innermost enclosure marks to the outermost marks.

Step 2. Exponents (Powers)

Step 3. $\times$, $\div$ (Compute in order from left to right, no matter which operation comes first)

Step 4. $+$, $-$ (Compute in order from left to right, no matter which operation comes first)

Note: ① $(a \times b) \div c = a \times (b \div c)$

$\because (a \times b) \div c = \dfrac{a \times b}{c} = \dfrac{ab}{c}$ and $a \times (b \div c) = a \times \dfrac{b}{c} = \dfrac{ab}{c}$

② $(a \div b) \times c \neq a \div (b \times c)$

③ $a \div b \times c = (a \div b) \times c \neq a \div (b \times c)$

for multiplying and dividing, calculate from left to right by order.

3. Multiplying Polynomials ○○○○

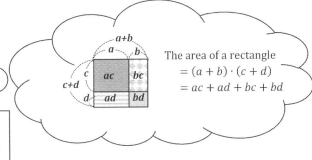

The area of a rectangle
$$= (a + b) \cdot (c + d)$$
$$= ac + ad + bc + bd$$

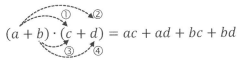

$$(a + b) \cdot (c + d) = ac + ad + bc + bd$$

Note

- $(a + b) \cdot (c + d) = a(c + d) + b(c + d) = ac + ad + bc + bd$

- $(x + a)(x + b) = x^2 + (a + b)x + ab$

- $(ax + b)(cx + d) = acx^2 + (ad + bc)x + bd$

- $(a + b + 1)(a + b - 1) = (A + 1)(A - 1), \quad Letting \ a + b = A$

$$= A^2 - A + A - 1 = A^2 - 1 = (a + b)^2 - 1$$

$$= (a + b)(a + b) - 1 = a^2 + ab + ba + b^2 - 1$$

$$= a^2 + 2ab + b^2 - 1$$

- $(a + 1)(a + 2)(a + 3)(a + 4) = (a + 1)(a + 4)(a + 2)(a + 3) \ ; \ Regroup \ to \ get \ a^2 + 5a$

$$= (a^2 + 5a + 4)(a^2 + 5a + 6)$$

$$= (A + 4)(A + 6), \quad Letting \ a^2 + 5a = A$$

$$= A^2 + 10A + 24$$

$$= (a^2 + 5a)^2 + 10(a^2 + 5a) + 24, \quad Replace \ A \ as \ a^2 + 5a$$

$$= a^4 + 10a^3 + 35a^2 + 50a + 24$$

> If the number of $(-)$ signs in the product is even, then $(+)$.
> $(-) \cdot (-) = (+) \ ; \quad (-) \cdot (-) \cdot (-) \cdot (-) = (+)$
>
> If the number of $(-)$ signs in the product is odd, then $(-)$.
> $(-) \cdot (+) = (-) \ ; \quad (-) \cdot (-) \cdot (-) = (-)$

4. Formulas

(1) $\boxed{(a + b)^2 = a^2 + 2ab + b^2}$

$\because (a + b)^2 = (a + b)(a + b) = a^2 + ab + ba + b^2 = a^2 + 2ab + b^2$

(2) $\boxed{(a - b)^2 = a^2 - 2ab + b^2}$

$\because (a - b)^2 = (a - b)(a - b) = a^2 - ab - ba + b^2 = a^2 - 2ab + b^2$

(3) $\boxed{a^2 + b^2 = (a + b)^2 - 2ab = (a - b)^2 + 2ab}$

(4) $\boxed{a^2 - b^2 = (a + b)(a - b)}$

$(a + b)^2 \neq a^2 + b^2$

$(a - b)^2 \neq a^2 - b^2$

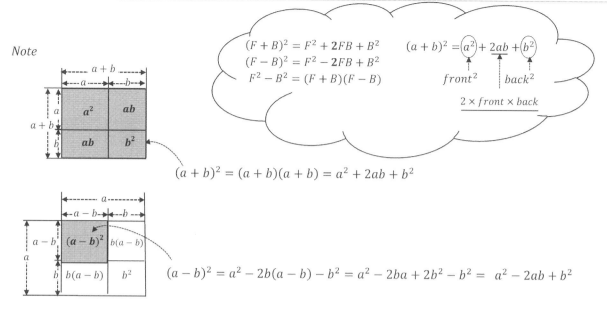

Note

$$(a + b)^2 = (a + b)(a + b) = a^2 + 2ab + b^2$$

$$(a - b)^2 = a^2 - 2b(a - b) - b^2 = a^2 - 2ba + 2b^2 - b^2 = a^2 - 2ab + b^2$$

Example $101^2 = (100 + 1)^2 = 100^2 + 2 \cdot 100 \cdot 1 + 1^2 = 10201$

$99^2 = (100 - 1)^2 = 100^2 - 2 \cdot 100 \cdot 1 + 1^2 = 9801$

$101 \cdot 99 = (100 + 1)(100 - 1) = 100^2 - 1^2 = 9999$

Expanding Formulas

(1) $\left(a + \dfrac{1}{a}\right)^2 = a^2 + 2 \cdot a \cdot \dfrac{1}{a} + \dfrac{1}{a^2} = a^2 + \dfrac{1}{a^2} + 2$

(2) $\left(a - \dfrac{1}{a}\right)^2 = a^2 - 2 \cdot a \cdot \dfrac{1}{a} + \dfrac{1}{a^2} = a^2 + \dfrac{1}{a^2} - 2$

(3) $a^2 + \dfrac{1}{a^2} = \left(a + \dfrac{1}{a}\right)^2 - 2 \cdot a \cdot \dfrac{1}{a} = \left(a + \dfrac{1}{a}\right)^2 - 2$

$\quad a^2 + \dfrac{1}{a^2} = \left(a - \dfrac{1}{a}\right)^2 + 2 \cdot a \cdot \dfrac{1}{a} = \left(a - \dfrac{1}{a}\right)^2 + 2$

(4) $a^2 - \dfrac{1}{a^2} = \left(a + \dfrac{1}{a}\right)\left(a - \dfrac{1}{a}\right)$

(5) $\left(a + \dfrac{1}{a}\right)^2 = \left(a - \dfrac{1}{a}\right)^2 + 4$

(6) $\left(a - \dfrac{1}{a}\right)^2 = \left(a + \dfrac{1}{a}\right)^2 - 4$

(7) $(a + b)^2 = (a - b)^2 + 4ab$

(8) $(a - b)^2 = (a + b)^2 - 4ab$

Note: $(a + b + c)^2 = ((a + b) + c)^2 = (A + c)^2,$ *Letting* $a + b = A$

$\qquad\qquad = A^2 + 2Ac + c^2$

$\qquad\qquad = (a + b)^2 + 2(a + b)c + c^2,$ *Replace A as $a + b$*

$\qquad\qquad = a^2 + 2ab + b^2 + 2ac + 2bc + c^2$

$\qquad\qquad = a^2 + b^2 + c^2 + 2ab + 2bc + 2ac,$ *Rearranging the terms*

5. Special Equalities

(1) The value of an algebraic expression $2x + 3y$ when $x = 3$ and $y = 4$ is

$(2 \cdot 3) + (3 \cdot 4) = 6 + 12 = 18$.

(2) When $x = y + 1$, $x + y = (y + 1) + y = 2y + 1$, replacing x with $y + 1$

(3) Solve an equation which contains two or more variables for a given variable by isolating the variable to one side of the equation.

Example

Solve the equation $ax + by = c$ for the variables x and y separately.

For x, $ax = -by + c$; $x = -\dfrac{b}{a}y + \dfrac{c}{a}$

For y, $by = -ax + c$; $y = -\dfrac{a}{b}x + \dfrac{c}{b}$

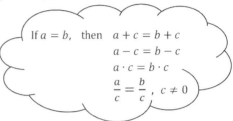

If $a = b$, then $a + c = b + c$
$a - c = b - c$
$a \cdot c = b \cdot c$
$\dfrac{a}{c} = \dfrac{b}{c}$, $c \neq 0$

Exercises

1. Simplify each expression.

(1) $a^2 \cdot a^3 \cdot a^4$

(2) $x^3 \cdot y^2 \cdot x^4 \cdot y \cdot z$

(3) $(2^3 xy^2 z^3)^2$

(4) $(x^3)^2 \cdot (x^4)^3$

(5) $((-x)^2)^3 \cdot ((-x)^3)^2$

(6) $(-a^2 b^3)^5$

(7) $-3xy^2 (-2x^2 yz^3)^3$

(8) $\left(-\dfrac{x}{y^2}\right)^2$

(9) $\dfrac{a^2 a^3}{(-a)^4}$

(10) $\left(\dfrac{2}{3}a^2\right)^2 \cdot \left(\dfrac{3}{4}a^3\right)^2$

(11) $(-a^2 b)^3 \div (-a)^3 \cdot (ab^2)^2$

(12) $\left(\dfrac{2}{3}\right)^{-3}$

(13) $\left(\dfrac{ab}{a^2 b^3}\right)^2$

(14) $\dfrac{x^3 x^{-4}}{x^2}$

(15) $\dfrac{a^3 b^{-2}}{a^{-4} b^3}$

(16) $\dfrac{2^3 + 2^3 + 2^3}{5^2 + 5^2 + 5^2}$

(17) $3^{2a-1} + 3^{2a-1} + 3^{2a-1}$

(18) $\dfrac{3^4 + 3^5 + 3^6 + 3^7}{3 + 3^2 + 3^3 + 3^4}$

(19) $\dfrac{4^3 + 4^3 + 4^3 + 4^3}{4^3 \cdot 4^3 \cdot 4^3 \cdot 4^3}$

(20) $-3xy^2 \cdot (-2x^2 y)^3 \div (2xy)^2$

(21) $\left(-\dfrac{3}{2}xy^3\right)^3 \div 4x^2 y \cdot \left(-\dfrac{4}{3}x^3 y\right)^2$

(22) $3^{-1} \cdot \left(\dfrac{1}{2}\right)^3 \cdot 3^3$

(23) $8^{a-1} \cdot 2^{3a+1} \div 4^{3a-1}$

2. Find all expressions that are true.

(1) $(a^2)^3 = a^5$

(2) $(-a)^3 \cdot -a^2 = -a^5$

(3) $a^3 \div a^3 = a^1$

(4) $a^4 \div a^3 \cdot a^5 = a^6$

(5) $a^2 + a^3 = a^5$

(6) $a^{-2} \cdot b^{-2} = (ab)^{-2}$

(7) $\left(\dfrac{a}{b^2}\right)^3 = \dfrac{a^3}{b^5}$

(8) $(a^2 b)^3 \div -2ab = -\dfrac{1}{2}a^5 b^2$

(9) $(a^2)^3 = a^{2^3}$

(10) $\left(\dfrac{3}{x}\right)^2 = \dfrac{1}{x^2}$

(11) $(2a^2)^3 = 6a^6$

3. Find a and b for the following:

(1) $32^3 = (2^a)^3 = 2^b$

(2) $2^{a+3} = 8^3$

(3) $(2^3)^2 \cdot (2^4)^a = 2^{18}$

(4) $(3^b)^3 \div 3^5 = 3^{10}$

(5) $(4^3)^a = 2^{42}$

(6) $24^4 = 2^a \cdot 3^b$

(7) $16^a = 2^{a+3}$

(8) $(2^3)^4 \div 8^3 \cdot (3^3)^2 = 2^a \cdot 3^b$

(9) $(2)^{3a+1} \div (2)^{2a-3} = 4$

(10) $5^a + 5^{a+2} = 3250$

(11) $2^a + 2^{a+2} = 160$

(12) $2^{a-2} = 0.5^{2a-1}$

(13) $(-2x^a)^b = -32x^{15}$

(14) $2^{a+2} = 2^{a+1} + 8$

(15) $(x^3y)^2 \cdot (xy^2)^a \div x^2y^3 = x^by^{13}$

4. $a = 2^{x+1}, b = 3^{x-1}$. Express 6^x using a and b.

5. $10^x = 2$, $10^y = 3$. Simplify $6^{\frac{x-y}{x+y}}$.

6. For a positive integer n, compute $(-1)^{2n+1} \cdot (-1)^{3n-1} \cdot (-1)^{2n-1} \div (-1)^{3n}$

.

7. Order the following numbers from least to greatest: 2^{32}, 4^{10}, 8^7, $\left(\frac{1}{2}\right)^{-30}$

8. Find the sum of all possible values of a natural number a which satisfies $a^{2a-1} = a^{3a-4}$.

9. For any positive number n, $2^{n+3}(3^{n+1} + 3^{n+2}) = a6^n$. Find the value of a.

10. For a solid with a length of a^3b^4, width of $3ab^2$, and volume of $15\,a^7b^8$, find the height.

11. Find the number of digits in the final value of the following expressions:

    (1) $2^4 \cdot 3^2 \cdot 5^5$

    (2) $5^2 \cdot 3^3 \cdot 20 \cdot 6$

    (3) $4^8 \cdot 5^{10}$

12. Simplify each polynomial.

    (1) $(2a + 3b) - (a - b)$

    (2) $(3a^2 - a + 3) - (-5a + 3)$

    (3) $(2a + 3) - (a^2 - 2a + 5)$

    (4) $4x - \{-2y + 3x - (2x - y) + 3\} - (x - 3y)$

    (5) $\frac{1}{3}x - \frac{2}{3}y - (2x + 3y) - \frac{1}{2}x$

    (6) $\frac{x+3y-1}{2} - \frac{2x-y+2}{3}$

    (7) $2a - [a^2 - \{3b - (2a - b) + a^2\} - 5]$

13. When an integer a is divided by 5, the remainder is 1.

When an integer b is divided by 5, the remainder is 2 . Find the remainder when $a + b$ is divided by 5.

14. a is the coefficient of x^2 and b is the constant of the following polynomials. Find $a - b$.

(1) $-(2x^2 - 4x + 5) + (3x^2 - x + 1)$

(2) $\left(\frac{1}{3}x^2 - \frac{1}{2}x + 2\right) - \left(\frac{1}{2}x^2 - \frac{2}{3}x + 5\right)$

(3) $2x^2 - \{3x - (3x^2 + 2x)\} - 2x + 5$

15. You wanted to add the polynomial $-2a^2 + 3a - 4$ to a polynomial A, but you accidentally subtracted the polynomial from A and got $-3a^2 + 5$. Compute the right answer.

16. If you subtract the polynomial $2a^2 - a + 3$ from two times a polynomial A, then you get $-2a^2 - a - 2$. If you add two times the polynomial $2a^2 - a + 3$ to a polynomial A, then you get $4a^2 + a + 2$. Find the value of a satisfying the two conditions.

17. Find the perimeter of the following shapes.

(1)

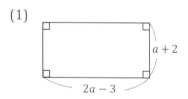

$a + 2$

$2a - 3$

(2)

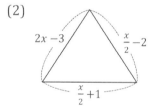

$2x - 3$

$\frac{x}{2} - 2$

$\frac{x}{2} + 1$

18. Simplify each polynomial.

(1) $-2x(3x + 4y - 2)$

(2) $x(x^2 - xy + y^2) - y(-x^2 + xy + y^2)$

(3) $(3a + 2b - 4ab) \cdot -\frac{1}{2}a$

(4) $(a^2b - ab^2) \div (-ab)$

(5) $(3a^2b - 2ab^2) \div \left(-\frac{2}{3}ab\right)$

(6) $2a(a - 1) - (a^2 - 1) - 2a(-a + 1)$

(7) $-\frac{5}{6}x^2y \cdot \left(\frac{3}{5}xy^2 - 3xy\right)$

(8) $\left(\frac{2}{3}xy^2 - 2x^2y^2\right) \cdot \left(-\frac{3}{2}xy\right) + \left(\frac{4}{3}x^2y - xy^2\right) \div \left(-\frac{3}{2}xy\right)$

(9) $\frac{4x^2 - 2xy}{2xy} - \frac{6xy^2 - 9y^2}{3xy}$

(10) $(6x^2y + 3x^2) \div 3x - (3xy - 9y^2) \div 3y$

(11) $\left(\frac{1}{8}ab - \frac{1}{2}a\right) \cdot 4b - \left(\frac{3}{4}a^2b^2 + a^2b\right) \div 3a$

(12) $\left\{\frac{1}{2}x^2 - \frac{2}{3}(x-3)\right\} + 3\left\{\frac{1}{2}(x-2) - \frac{1}{3}(x^2+3) + 2\right\}$

 # 19. $(2x^2y^3)^a \div 4xy \cdot \frac{1}{2}x^2y = bx^3y^3$. Find the value of $a + b$, where a and b are constants.

 # 20. Find the value of $a + b + c$ for the following, where $a, b,$ and c are constants:

(1) $\left(\frac{4}{3}x^2y - 3xy^2 + 2xy\right) \div \frac{1}{2}xy = ax + by + c$

(2) $\frac{1}{2}(x^2 - 3x + 1) - 2x(x - 1) + 3(4x^2 - 3x - 2) = ax^2 + bx + c$

 # 21. Find the polynomial for each expression.

(1) If a polynomial is multiplied by $2ab$, the result is $\frac{1}{2}a^2b + ab^2 - \frac{1}{3}ab$.

(2) If a polynomial is divided by $3a - 2b$, the quotient is $\frac{1}{4}ab$ and there is no remainder.

 # 22. Find the area of the shaded part in the rectangle.

The rectangle has a length of $4a$ and width of $2b$.

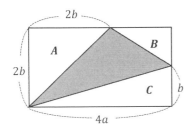

#23. Expand and simplify each polynomial.

(1) $(2x - 5)(x + 3)$

(2) $(2x - 1)(3x^2 - x - 2)$

(3) $\left(x + \frac{1}{3}\right)\left(x - \frac{1}{2}\right)$

(4) $(3 - 2a)(3 + 2a)$

(5) $(3a - 2b)(3a + 2b)$

(6) $(2x + 3)(2x + 3)$

(7) $(-2x + 3)(-2x - 3)$

(8) $(a^3 + b^3)(a^3 - b^3)$

(9) $\left(-4x - \frac{1}{2}\right)^2$

(10) $(x + y - 2)^2$

(13) 99^2

(14) $(x + 2y + 3z)(x + 2y - 3z)$

(15) $(2x + y - 3)(2x - y + 3)$

(11) 102×98

(12) 92×93

(16) $111 \times 109 - 107 \times 113$

(17) $(2a + b)^2 - (2a - b)^2$

(18) $(a - 3)(a + 2)(a - 1)(a + 4)$

24. Find the area of the shaded part of each shape.

(1)

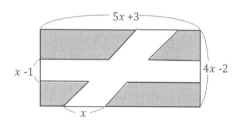

(2)

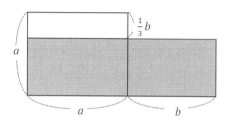

(3)

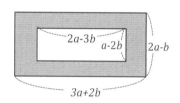

25. Evaluate the polynomial for the variable in each expression.

(1) $3x - 2$ for $x = -2$

(2) $\frac{2}{3}x + 3$ for $x = -1$

(3) $-2x^2 - 3x + 1$ for $x = -3$

(4) $(2x - 2)(-3x + 1)$ for $x = 2$

26. Find the values of the following polynomials:

(1) $a^2 + \dfrac{1}{b^2}$ when $a - \dfrac{1}{b} = 3, \ \dfrac{b}{a} = -\dfrac{1}{3}$

(2) $a^2 + \dfrac{1}{a^2}$ when $a + \dfrac{1}{a} = 3$

(3) ab when $a - b = 4, \ a^2 + b^2 = 8$

(4) $\left(a - \dfrac{1}{a}\right)^2$ when $a + \dfrac{1}{a} = -3$

(5) $\dfrac{3a+3b}{2a-4b}$ when $\dfrac{a-b}{a+b} = \dfrac{2}{3}$

(6) $\dfrac{b-c}{a} + \dfrac{c-a}{b} - \dfrac{a+b}{c}$ when $a + b - c = 0, \ (abc \neq 0)$

(7) $a^4 + b^4$ when $a - b = 1, \ ab = 2$

(8) $\dfrac{(3a-2b)^2}{(2a+3b)^2}$ when $a : b = 3 : 2$

(9) $\left(\dfrac{2}{3}a^2 - \dfrac{3}{2}b^2\right)\left(-\dfrac{2}{3}a^2 - \dfrac{3}{2}b^2\right)$ when $a = \dfrac{1}{2}, \ b = \dfrac{1}{3}$

(10) $\dfrac{-3a-6ab+3b}{a-3ab-b}$ when $\dfrac{1}{a} - \dfrac{1}{b} = 3, \ ab \neq 0$

(11) $x^2 + \dfrac{9}{x^2} - 3$ when $x^2 - 4x - 3 = 0$

(12) $x^2 + \dfrac{1}{x^2} - 2x + \dfrac{2}{x}$ when $x^2 + 3x - 1 = 0$

(13) $\dfrac{y}{x} + \dfrac{x}{y}$ when $x - y = 3, \ (x+2)(y-2) = -6$

(14) $\dfrac{3x+4xy-3y}{x-y}$ when $\dfrac{1}{x} - \dfrac{1}{y} = 2$

(15) $\dfrac{x-2y}{2x+y}$ when $\dfrac{3x+y}{2} = \dfrac{2x-y}{3}$

(16) $\dfrac{a^2-b^2}{(a+b)^2}$ when $(3x+a)(bx-1) = (3x-1)^2$

(17) $\dfrac{1}{xyz}$ when $x + \dfrac{1}{y} = 1, \ y + \dfrac{1}{z} = 1$

(18) $(x+1)(x+2)(x-3)(x-4)$ when $x^2 - 2x - 5 = 0$

27. Evaluate each equation for the specified variable.

(1) $2x - 3y + 6 = 0$ for x

(2) $x = -2y + 3$ for y

(3) $2a = \dfrac{1}{3}(2b - 1)$ for b

(4) $c = \dfrac{5}{9}(F - 32)$ for F

(5) $a + b : a - b = 3 : 5$ for a

(6) $\dfrac{1}{a} + \dfrac{1}{b} = \dfrac{1}{c}$, $(a \neq 0, b \neq 0, c \neq 0)$ for a

(7) $(2a - b)(a + b) = (a + 3b)(2a - b)$ for a

(8) $a = -\dfrac{c}{b} + 1$, $b \neq 0$, $c \neq 0$ for b

28. Find the value of $a + b$, where a and b are constants.

(1) $(5 - 1)(5 + 1)(5^2 + 1)(5^4 + 1) = 5^a - b$

(2) $8(3^2 + 1)(3^4 + 1)(3^8 + 1) = 3^a + b$

(3) $(2x + ay)(x - 3y) = 2x^2 - bxy + 9y^2$

(4) $(x + y)(x - y) - (3y - 2x)(2x - 3y) = \dfrac{1}{a}x^2 - 12xy - \dfrac{1}{b}y^2$

29. Find the sum of the coefficients as well as the constants for each polynomial.

(1) $(x + 2y - 3)(x + 2y - 2)$

(2) $(2x - 3y - 5)(2x + 3y - 5)$

(3) $(3x + 2y)(x - 2y) - (x + 3y)(x - 2y)$

Chapter 8. Systems of Equations

8-1 Systems of Equations

1. Definition

A linear equation is an equation in the form of

$$ax + by = c$$, where $a, b,$ and c are constants and $a \neq 0, b \neq 0$.

For example, $2x + 3y - 5 = 0$ or $2x + 3y = 5$.

A system of equations is a pair of two or more linear equations with the same variables.

For example, $\begin{cases} x + y = 2 \\ 2x + y = 3 \end{cases}$

For the values of x and y which are satisfying the two linear equations in the system, the ordered pair (x, y) is called the *solution* for the system of equations.

Note: $A = Set\ of\ solutions\ for\ equation$ ①
       $B = Set\ of\ solutions\ for\ equation$ ②.

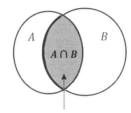

Solutions for the system of equations ① and ②.

Example

For any positive integers x and y, find the solution for the system $\begin{cases} x + y = 3 \cdots\cdots ① \\ 2x + y = 5 \cdots\cdots ② \end{cases}$

For equation ①, the solutions are $(1, 2), (2, 1)$ and for an equation ②, the solutions are $(1, 3), (2, 1)$.

Therefore, the solution for the system of equations ① and ② is $(2, 1)$.

2. Solving Systems of Equations

(1) The Elimination Method

The *elimination method* is a method for finding a solution by eliminating (removing) one variable by using addition or subtraction.

Note 1: If the sign of the coefficient of one variable which is supposed to be removed is different from the sign of the coefficient of the other variable, use addition for the two equations to remove the variable.

If the sign of the coefficient of one variable which is supposed to be removed is the same as the sign of the coefficient of the other variable, use subtraction for the two equations to remove the variable.

Note 2: One or two of the given equations needs to be multiplied by a number in order to make the two equations with the same absolute value for the coefficient of the variable which is supposed to be removed.

Example
$$\begin{cases} 2x + y = 3 & \cdots\cdots ① \\ x - y = 6 & \cdots\cdots ② \end{cases}$$

Step 1.   Add equations ① and ② to remove the variable y. Then solve for x.

$$\begin{array}{r} 2x + y = 3 \cdots\cdots ① \\ +)\ \underline{x - y = 6 \cdots\cdots ②} \\ 3x \quad\ = 9\ ;\ x = 3 \end{array}$$

Step 2.   Substitute the value of x into any one of the given two equations and then solve for y.

$$x - y = 6 \ \Rightarrow\ 3 - y = 6 \ \Rightarrow\ y = -3$$

Step 3.   Find the solution.

$$(x, y) = (3, -3)$$

Example
$$\begin{cases} 2x + y = 4 & \cdots\cdots ① \\ 3x + y = 2 & \cdots\cdots ② \end{cases}$$

Step 1.   Subtract equation ② from equation① to remove the variable y.

Then solve for x.

$$\begin{array}{r} 2x + y = 4 \cdots\cdots ① \\ -)\ \underline{3x + y = 2 \cdots\cdots ②} \\ -x \quad\ = 2 \ \ ;\ x = -2 \end{array}$$

Step 2.   Substitute the value of x into any one of the given two equations.

Then solve for y.

$$2x + y = 4 \ \Rightarrow\ -4 + y = 4 \ \Rightarrow\ y = 8$$

Step 3.   Find the solution.

$$(x, y) = (-2,\ 8)$$

Example
$$\begin{cases} 2x + 3y = 3 \cdots\cdots ① \\ 3x + 4y = 1 \cdots\cdots ② \end{cases}$$

Step 1.   Consider *Note*1 and *Note* 2.

$$6x + 9y = 9 \cdots\cdots ① \times 3$$
$$-)\ \underline{6x + 8y = 2} \cdots\cdots ② \times 2$$
$$y = 7$$

Step 2.   Substitute the value of y into any one of the given two equations.

Then solve for x.

$$2x + 3y = 3 \Rightarrow 2x + 21 = 3 \Rightarrow 2x = -18 \Rightarrow x = -9$$

Step 3.   Find the solution.

$$(x, y) = (-9,\ 7)$$

(2) The Substitution Method

The substitution method is a method for finding the solution by substituting the expression which is already solved for one variable into the other equation.

Note: *If the coefficient of one variable is 1 or −1 or if one of the two equations is easily solved for one variable, – for example, (x = expression of y) or (y = expression of x) − this substitution method is the best way to find the solution.*

Example
$$\begin{cases} 2x + y\ = 3 \cdots\cdots ① \\ 3x + 2y = 5 \cdots\cdots ② \end{cases}$$

Step 1.   Solve equation ① for one variable y in terms of the other; $y = -2x + 3$

Step 2.   Substitute the expression $y = -2x + 3$ found in step 1 into the other equation ② to obtain an equation in one variable.

$$3x + 2(-2x + 3) = 5\ ;\ -x = -1\ ;\ x = 1$$

Step 3.   Substitute $x = 1$ into equation ① to find the value of the other variable;

$$2 \cdot 1 + y = 3\ ;\ y = 1$$

Step 4.   Find the solution.

$$(x, y) = (1, 1)$$

Step 5.   Check the solution to see that it satisfies each of the given equations.

Note: If the coefficients of variables are unknown and the solution for the system of equations is given, then do the following:

Step 1. Substitute the solution into the given system of equations.

Step 2. Find the coefficients of variables using the elimination method or substitution method.

Example
$$\begin{cases} ax + by = 3 & \cdots\cdots ① \\ 2bx - ay = -4 & \cdots\cdots ② \end{cases}$$

The solution for this system of equations is $(1, -2)$. Find the value of $a + b$.

Step 1.
$$\begin{cases} a - 2b = 3 & \cdots\cdots ③, \quad \text{from } ① \\ 2b + 2a = -4 & \cdots\cdots ④, \quad \text{from } ② \end{cases}$$

Step 2.
$$\begin{array}{rl} a - 2b &= 3 \\ +)\ 2a + 2b &= -4 \\ \hline 3a &= -1 \quad ; \ a = -\frac{1}{3} \end{array}$$

Substituting $a = -\frac{1}{3}$ into ③, $b = -\frac{5}{3}$

Therefore, $a + b = -\frac{1}{3} - \frac{5}{3} = -2$

3. Solving Special Systems

(1) If the coefficients of variables are fractions or decimals

⇒ Change the coefficients to integers. Then solve as usual.

Example Solve a system
$$\begin{cases} \frac{1}{2}x - \frac{1}{3}y = 1 & \cdots\cdots ① \\ 0.3x + 0.2y = 0.2 & \cdots\cdots ② \end{cases}$$

Step 1.
$$\begin{cases} 3x - 2y = 6 & \cdots\cdots ① \times 6 \\ 3x + 2y = 2 & \cdots\cdots ② \times 10 \end{cases}$$

Step 2.
$$\begin{array}{rl} 3x - 2y &= 6 \cdots\cdots ③ \\ -)\ 3x + 2y &= 2 \cdots\cdots ④ \\ \hline -4y &= 4 \quad ; \ y = -1 \end{array}$$

Substituting $y = -1$ into ③, $x = \frac{4}{3}$

Therefore, $(x, y) = \left(\frac{4}{3}, -1\right)$

$(2)\begin{cases} px = qy + r \\ my = nx + t \end{cases} \Rightarrow$ Rearrange or simplify in the form of $ax + by = c$

Example $\begin{cases} 2x = 3y + 5 \\ -y = 4x - 3 \end{cases} \xrightarrow{\text{Rearrange}} \begin{cases} 2x - 3y = 5 \ \cdots\cdots ① \\ 4x + y = 3 \ \cdots\cdots ② \end{cases}$

$\begin{cases} 4x - 6y = 10 \ \cdots\cdots ① \times 2 \\ 4x + y = 3 \cdots\cdots ② \end{cases}$

$4x - 6y = 10 \ \cdots\cdots ③$

$-)\ \underline{4x + y = 3 \ \cdots\cdots ④}$

$\qquad -7y = 7 \ \ ; y = -1$

Substituting $y = -1$ into ④, $x = 1$

Therefore, $(x, y) = (1, -1)$

$(3)\begin{cases} \dfrac{a}{x} + \dfrac{b}{y} = c \\ \dfrac{m}{x} + \dfrac{n}{y} = t \end{cases} \Rightarrow$ Let $\dfrac{1}{x} = A, \ \dfrac{1}{y} = B$

Example $\begin{cases} \dfrac{2}{x} + \dfrac{3}{y} = 1 \\ \dfrac{1}{x} + \dfrac{2}{y} = 2 \end{cases} \xrightarrow{\text{Replace}} \begin{cases} 2A + 3B = 1 \ \cdots\cdots ① \\ A + 2B = 2 \ \cdots\cdots ② \end{cases}$

$\begin{cases} 2A + 3B = 1 \ \cdots\cdots ① \\ 2A + 4B = 4 \cdots\cdots ② \times 2 \end{cases}$

$2A + 3B = 1 \ \cdots\cdots ①$

$-)\ \underline{2A + 4B = 4 \ \cdots\cdots ③}$

$\qquad -B = -3 \ \ ; B = 3 \ ; \dfrac{1}{y} = 3 \ ; y = \dfrac{1}{3}$

Substituting $B = 3$ into ②, $A = -4 \ ; \dfrac{1}{x} = -4 \ ; x = -\dfrac{1}{4}$

Therefore, $(x, y) = \left(-\dfrac{1}{4}, \dfrac{1}{3}\right)$

(4) A system of equations in form of the $A = B = C$

$\Rightarrow$ Rewrite the equations as a system

$\begin{cases} A = B \\ B = C \end{cases}$ or $\begin{cases} A = B \\ A = C \end{cases}$ or $\begin{cases} A = C \\ B = C \end{cases}$

Note : If $A = B = k$, k is a constant, rewrite the equation as a system $\begin{cases} A = k \\ B = k \end{cases}$.

Example   Solve the system $2x - y = x + 3y = y + 5$.

Step 1. (Rewrite)

$$\begin{cases} 2x - y = x + 3y & \cdots\cdots ① \\ 2x - y = y + 5 & \cdots\cdots ② \end{cases}$$

Step 2. (Rearranging ① and ②)

$$\begin{cases} x - 4y = 0 & \cdots\cdots ③ \\ 2x - 2y = 5 & \cdots\cdots ④ \end{cases}$$

Step 3. (Solve)

$$2x - 8y = 0 \quad\cdots\cdots ③ \times 2$$
$$-)\ \underline{2x - 2y = 5 \quad\cdots\cdots ④}$$
$$-6y = -5 \quad ; \ y = \frac{5}{6}$$

Substituting $y = \frac{5}{6}$ into ③, $x = \frac{10}{3}$

Therefore, $(x, y) = \left(\frac{10}{3}, \frac{5}{6} \right)$

---

(5) A system of three equations with three variables

$\Rightarrow$ Step 1. Eliminate one of the variables to create a system of equations with two variables.

Step 2. Solve the new system for two variables.

Step 3. Substitute the two values to find the eliminated variable.

Step 4. Find the solution.

Example   Find the solution (x, y, z) for the system $\begin{cases} 2x - y + z = 30 & \cdots\cdots ① \\ x + 2y + z = 2 & \cdots\cdots ② \\ 3x + 2y - z = -18 & \cdots\cdots ③ \end{cases}$

Step 1. (To remove z)

$$\begin{cases} 5x + y = 12 & \cdots\cdots ④ \ (\text{using } ① + ③) \\ 4x + 4y = -16 & \cdots\cdots ⑤ \ (\text{using } ② + ③) \end{cases}$$

Step 2. (To find (x, y))

$$20x + 4y = 48 \ (\text{using } ④ \times 4)$$
$$-)\ \underline{4x + 4y = -16}$$
$$16x \quad\ = 64 \quad ; \ x = 4$$

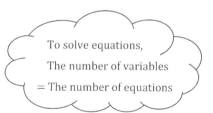

To solve equations,

The number of variables

= The number of equations

Substituting $x = 4$ into ④, $y = 12 - 20 = -8$

So, $(x, y) = (4, -8)$

Step 3. (To find z)

Substituting $(x, y) = (4, -8)$ into ①,

$$z = 30 - 2x + y = 30 - 2 \cdot 4 - 8 = 14$$

Step 4. $(x, y, z) = (4, -8, 14)$

Note: $\begin{cases} ax + by = m \cdots\cdots ① \\ by + cz = n \cdots\cdots ② \\ cz + ax = t \cdots\cdots ③ \end{cases}$ $\xrightarrow{\text{Add all three equations}}$ $2(ax + by + cz) = m + n + t$

So, $ax + by + cz = \frac{1}{2}(m + n + t)$

Using ①, $cz = \frac{1}{2}(m + n + t) - m$

Using ②, $ax = \frac{1}{2}(m + n + t) - n$

Using ③, $by = \frac{1}{2}(m + n + t) - t$

At this point, it's easy to find the solution (x, y, z).

---

(6) If a system $\begin{cases} ax + by = c \\ mx + ny = t \end{cases}$ has the special condition $\frac{a}{m} = \frac{b}{n} = \frac{c}{t}$, then

the system has an unlimited number of solutions. (Consistent and dependent)

Example $\begin{cases} 2x + 3y = 5 \\ 4x + 6y = 10 \end{cases}$

$\Rightarrow \dfrac{2}{4} = \dfrac{3}{6} = \dfrac{5}{10}$ ∴ The system has unlimited solutions.

> If the two equations are the same (in that they have the same ratios), that is, form of $0 = 0$, then it's always true.

> If the two equations have the same coefficients for the variables but constants, that is, form of $0 = a$ (a is a non-zero number), then it's impossible.

---

(7) If a system $\begin{cases} ax + by = c \\ mx + ny = t \end{cases}$ has the special condition $\frac{a}{m} = \frac{b}{n} \neq \frac{c}{t}$, then

the system does not have a solution. (Inconsistent)

> $ax + by + c = 0$ $\xrightarrow{\text{Transform}}$ $y = px + q$ (Equation $\xrightarrow{\text{Transform}}$ Function)
>
> If $y_1 = px + q_1$ and $y_2 = px + q_2$, $q_1 \neq q_2$, then y_1 and y_2 are parallel.
>
> So, the lines do not intersect at all.
>
> Therefore, $\dfrac{a}{m} = \dfrac{b}{n} \neq \dfrac{c}{t}$ $\Rightarrow$ The system has no solution.

Example $\begin{cases} 3x + y = 2 \\ 6x + 2y = 3 \end{cases}$

$\Rightarrow \dfrac{3}{6} = \dfrac{1}{2} \neq \dfrac{2}{3}$ $\therefore$ The system has no solution.

4. Steps for Solving Word Problems (with Two Variables)

Step 1. Assign two variables to solve the unknowns.

Step 2. Create a system of equations and solve it, using any applicable methods.

Step 3. Check your answer.

5. Graphing a System of Equations

For any constants $a, b,$ and $c,$ a system $\begin{cases} ax + by = c \\ a'x + b'y = c' \end{cases}$ in a coordinate plane creates one of

three types of cases : parallel, intersecting at one point, or coinciding.

To graph a system, find the x-intercept and y-intercept for each equation. Then graph both

equations on the same coordinate plane.

The solution of a system of equations will be the intersection point of the two lines.

Note : x -intercept is the x-coordinate of the point where the line crosses (intersects) the x-axis.
(x-intercept is the value of x when y = 0.)

y -intercept is the y-coordinate of the point where the line crosses (intersects) the y-axis.
(y-intercept is the value of x when x = 0.)

Example   Solve the system $\begin{cases} 2x + y = 4 \\ 4x - y = 2 \end{cases}$ by graphing.

$2x + y = 4 \Rightarrow y = -2x + 4$ $\therefore x$-intercept: $(2, 0)$ and y-intercept: $(0, 4)$

$4x - y = 2 \Rightarrow y = 4x - 2$ $\therefore x$-intercept: $\left(\dfrac{1}{2}, 0\right)$ and y-intercept: $(0, -2)$

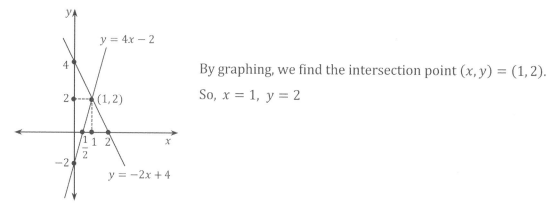

By graphing, we find the intersection point $(x, y) = (1, 2)$.

So, $x = 1, \ y = 2$

Exercises

#1. Solve each system.

(1) $\begin{cases} x + y = 4 \\ 2x - y = 5 \end{cases}$

(2) $\begin{cases} 2x + y = 3 \\ 2x + 3y = 4 \end{cases}$

(3) $\begin{cases} 2x + y = 7 \\ 3x + 2y = 5 \end{cases}$

(4) $\begin{cases} x - y = 5 \\ 2x + 5y = 3 \end{cases}$

(5) $\begin{cases} 3x - y = 0 \\ 5x - 2y = -3 \end{cases}$

(6) $\begin{cases} 4x - 3y = 6 \\ 6x - 2y = -6 \end{cases}$

#2. Find the value of ab for each system.

(1) $\begin{cases} ax - by = -2 \\ bx + 2y = a \end{cases}$ with solution $(3,2)$

(2) $\begin{cases} x + 5y = -3 \\ 2x - by = 5 \end{cases}$ with solution $(a, -1)$

(3) $\begin{cases} -2x + y = 5 \\ x - 2y = -1 \end{cases}$ with solution (a, b)

(4) $\begin{cases} 3x - by = 2 \\ ax + y = -2 \end{cases}$ with solution $(b - 1, 2)$

#3. The system $\begin{cases} 2x - y = 3 \\ x + 3y = 5 \end{cases}$ has a solution $(a + 1, b - 1)$. Find the value of $(a + b)^2 - (a - b)^2$.

#4. The system $\begin{cases} 2x + 3y = 5 \\ -x - 2y = -3 \end{cases}$ has a solution (a, b).

Find the solution for the system $\begin{cases} (3 - a)x + 2y = -2 \\ 2x + 3y = 2b + 1 \end{cases}$.

#5. The solution of the system $\begin{cases} 3x - 2y = -2 \\ (k - 1)x + y = -3 \end{cases}$ is the same as the solution of the equation

$2x - y = 3$. Find the constant k.

#6. Two systems $\begin{cases} 2x + by = 4 \\ x + 2y = -3 \end{cases}$ and $\begin{cases} x - 3y = 2 \\ ax + 2y = -1 \end{cases}$ have the same solution.

Find the value of $a + b$.

#7. Solve each system.

(1) $\begin{cases} x = 2y + 1 \\ x - y = 3 \end{cases}$

(2) $\begin{cases} y = x - 3 \\ 2x - y = 2 \end{cases}$

(3) $\begin{cases} \dfrac{1}{2}x + \dfrac{1}{3}y = 2 \\ \dfrac{2}{3}x - \dfrac{3}{4}y = -\dfrac{11}{12} \end{cases}$

(4) $\begin{cases} \dfrac{2}{x} + \dfrac{3}{y} = 4 \\ \dfrac{1}{2x} - \dfrac{1}{y} = 2 \end{cases}$

(5) $\begin{cases} 0.2x - 0.5y = 0.25 \\ -0.3x + 0.4y = -0.2 \end{cases}$

(6) $\begin{cases} 3x - y = -5 \\ 6x - 2y = 3 \end{cases}$

(7) $\begin{cases} x = \dfrac{1}{2}y + 1 \\ 2x - 4y = -4 \end{cases}$

(8) $\begin{cases} -\dfrac{2}{3}x + \dfrac{1}{4}y = -2 \\ 2x - \dfrac{3}{4}y = 6 \end{cases}$

(9) $\begin{cases} \dfrac{2}{x} + \dfrac{2}{y} = 1 \\ \dfrac{1}{x} - \dfrac{1}{y} = -1 \end{cases}$

(10) $\dfrac{x+1}{2} + \dfrac{y-1}{3} = \dfrac{2x-1}{3} + \dfrac{y+2}{4} = 2x + \dfrac{y}{2}$

(11) $\begin{cases} 2x - 3y = -4 \\ -3y + z = 2 \\ z + 2x = -6 \end{cases}$

(12) $\begin{cases} x : y + 1 = 2 : 3 \\ 3 : y - 1 = 4 : x - 1 \end{cases}$

#8. The system $\begin{cases} \dfrac{a+1}{2}x - \dfrac{3}{4}y = -2 \\ 5x + \dfrac{b-1}{2}y = 4 \end{cases}$ has an unlimited number of solutions.

Find the value of $a + b$.

#9. The system $\begin{cases} a(x - y) + \dfrac{y}{2} = -1 \\ -\dfrac{x}{2} - \dfrac{1}{a}y = 3 \end{cases}$ has no solution. Find the value of a.

#10. The system $\begin{cases} 2kx - (3x + y) = 2y \\ -(k - 1)x + 2y = kx \end{cases}$ has a solution other than (0,0).

Find the value of the constant k.

#11. Find the value of $a + b$ for the following systems:

(1) $\begin{cases} 2x - \dfrac{y}{3} = \dfrac{2}{3} \\ (x - y) : 3 = -1 : 1 \end{cases}$ with solution $(a, b - 1)$.

(2) $\begin{cases} \dfrac{3}{x} + \dfrac{2}{y} = b \\ \dfrac{1}{x} + \dfrac{2}{y} = \dfrac{1}{2} \end{cases}$ with solution $(a, 2a)$.

(3) $ax + (b - 1)y = 2ax - 3y + 5 = x + by - 1$ with solution $(2, 3)$.

#12. The system $\begin{cases} 3x - 2y = k \\ -2x + y = 3 \end{cases}$ has the solution (a, b) with the condition $a : b = 1 : 3$

Find the constant k.

#13. Find the value of $x + y$ for variables x and y that satisfy the equations $2^x \cdot 8^y = 32$ and $3^{x+1} \cdot 9^{y-1} = 3^3$

#14. Find the value of $\dfrac{1}{x} - \dfrac{1}{y}$ for variables x and y that satisfy the system

$\begin{cases} 2x - xy - 2y - 3 = 0 \\ 3x + 2xy - 3y + 1 = 0 \end{cases}.$

#15. The perimeter of a rectangle is 18 inches. The length of the rectangle is 3 inches shorter than twice its width. What is the area of the rectangle?

#16. Movie ticket prices are $6 for children and $9 for adults. Nichole pays $84 for 12 people. How many children are in her group?

#17. Apples and peaches are mixed in a box. There are 3 less apples than three times the number of peaches. Two times the total number of apples and peaches is 10. How many apples and peaches are in the box?

#18. Richard prepares a bag of candies for kids. If each kid gets 8 candies, then 8 candies will be left. If they each get 10 candies then Richard will be short 6 candies. How many candies are in Richard's bag?

#19. Nichole has quarters and dimes worth $2.55 in her purse. The number of dimes is two less than three times the number of quarters. How many quarters and dimes are in her purse?

#20. If you add the ten's digit and the one's digit of a certain two-digit integer, the sum is 12. If the digits of the number are interchanged, the new number will be 12 less than twice the original number. Find the original number.

#21. If 30 ounces of salt solution containing a x% of salt solution is added to 40 ounces of salt solution containing a y% of salt solution, it produces a salt solution that is 15% salt.
If 30 ounces of a salt solution containing a y% of salt solution is added to 40 ounces of a salt solution containing a x% of salt solution, it produces a salt solution that is 18% salt.
Find the values of x and y.

#22. Richard wants to produce 70 ounces of salt solution that is 8% salt by adding water after mixing salt solution that is 5% salt with salt solution that is 10% salt. The amount of a10% of salt solution is three times as much as the amount of a 5% of salt solution. How much water should he add?

#23. Nichole wants to produce 29 liters of 20% alcohol solution by adding alcohol after mixing two alcohol solutions that are 4% alcohol and 3% alcohol separately. The amount of alcohol solution that is 3% alcohol is twice the amount of alcohol solution that is 4% alcohol. How many liters of alcohol must be added?

#24. Nichole started to run at 9:50AM at a speed of 8 miles per hour and then walked the rest of the way at 3 miles per hour. She arrived at 10:40AM. If Nichole went to the park which was 4 miles away, how many miles did she run?

#25. Richard starts a trail ride in a parking lot. He rides up a long hill on A trail at 4 miles per hour and comes down the hill on B trail going 12 miles per hour. His ride takes 1 hour 20 minutes total. The total distance of A trail and B trail is 10 miles. How many miles long is B trail?

#26. Richard and Nichole want to finish a job. Richard works alone for 3 hours and leaves. Nichole comes and works alone the rest of the job for another 3 hours, thereby finishing the job.
OR if Richard works alone for 6 hours and then Nichole works alone for 2 hours for the rest of the job after Richard is done, the job is also completed. How long will it take Richard to finish the job by himself the entire time?

#27. 5 years ago, Nichole was 5 years less than one-third her mom's age. In 6 years, her mom will be 10 years more than twice Nichole's age at that time. How old was Nichole's mom when Nichole was 15?

#28. Richard walks from home to a library at 3 miles per hour. 20 minutes after Richard leaves, Nichole rides a bike at 8 miles per hour from home to the library. They arrive at the same time. How long does it take Richard to meet Nichole at the library?

#29. If Richard drives a car from home to the doctor's office at 50 miles per hour, he will arrive at the office 5 minutes earlier than his appointment time. If he drives a car at 40 miles per hour on the same route, he will arrive 10 minutes late to his appointment. How far is the office from Richard's home?

#30. Richard and Nichole jog towards each other from two opposite starting points 1 mile apart. Nichole jogs 1.5 times faster than Richard. How fast does Nichole have to jog if they want to meet each other in 30 minutes?

#31. There were 44 boys and girls in a math club last summer. This year, 25% of the boys quit and 15% of the girls joined the club again. Now the club has 41 members. How many boys and girls are in the club now?

#32. Six years ago, Nichole was three times as old as Richard. Four years from now, Nichole will be twice as old as Richard. How old is Richard now?

#33. Solve each system by graphing.

(1) $\begin{cases} x + y = 4 \\ 2x - y = 5 \end{cases}$
(2) $\begin{cases} 2x - y = 5 \\ 4x - 2y = 6 \end{cases}$
(3) $\begin{cases} 3x + 4y = 5 \\ 6x + 8y = 10 \end{cases}$

Chapter 9. Systems of Inequalities

9-1 Systems of Inequalities

1. Definition

A *system of inequalities* is a pair of two or more linear inequalities.

The *solution* for a system of inequalities is the value or range which satisfies both inequalities at the same time.

2. Solving Systems of Inequalities

Step 1. Solve each inequality and find the solutions for each.

Step 2. Graph the solutions on a number line.

Step 3. Find the common range or value from the solutions of all inequalities.

When $a < b$,

(1) $\begin{cases} x < a \\ x < b \end{cases} \Rightarrow x < a$

(2) $\begin{cases} x > a \\ x > b \end{cases} \Rightarrow x > b$

(3) $\begin{cases} x > a \\ x < b \end{cases} \Rightarrow a < x < b$

3. Solving Special Systems

(1) Solve parentheses by using the distributive property.

(2) Change fraction or decimal coefficients of variables to integers.

(3) Rewrite the system $A < B < C$ as $\begin{cases} A < B \\ B < C \end{cases}$.

$A = B = C \Rightarrow$

$\begin{cases} A = B \\ B = C \end{cases}$ or $\begin{cases} A = B \\ A = C \end{cases}$ or $\begin{cases} A = C \\ B = C \end{cases}$

But, $A < B < C \Rightarrow \begin{cases} A < B \\ B < C \end{cases}$ only

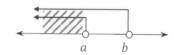

$\begin{cases} A < B \\ A < C \end{cases}$ ($\because$ We can't compare B and C)

$\begin{cases} A < C \\ B < C \end{cases}$ ($\because$ We can't compare A and B)

Don't rewrite the system when only B has an unknown variable.

For example,

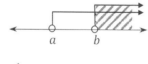

$\Rightarrow 1 < x < \frac{3}{2}$

Constant ... x term ... Constant

(4) $\begin{cases} x \le a \\ x \ge a \end{cases}$ $\Rightarrow$ $x = a$ (This is the only solution.)

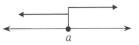

(5) If there is no common value or range,

$\begin{cases} x < a \\ x \ge a \end{cases}$ $\qquad$ $\begin{cases} x < a \\ x > a \end{cases}$ $\qquad$ $\begin{cases} x < a \\ x \ge b \end{cases}$, $a < b$

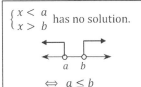

$\Rightarrow$ No solution $\qquad$ $\Rightarrow$ No solution $\qquad$ $\Rightarrow$ No solution

(6) Conditions for having solutions:

1) $\begin{cases} x \le a \\ x \ge b \end{cases}$  $\Rightarrow b \le a$

$(\because$ If $b = a$, then $\begin{cases} x \le a \\ x \ge a \end{cases}$ $\therefore x = a$ is the solution.$)$

2) $\begin{cases} x \le a \\ x > b \end{cases}$ $\Rightarrow b < a$

$(\because$ If $b = a$, then $\begin{cases} x \le a \\ x > a \end{cases}$ $\therefore$ There is no solution. Therefore, $b \lneqq a$ $(b < a)$. $)$

3) $\begin{cases} x < a \\ x \ge b \end{cases}$ 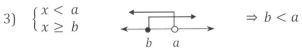 $\Rightarrow b < a$

$(\because$ If $b = a$, then $\begin{cases} x < a \\ x \ge a \end{cases}$ $\therefore$ There is no solution. Therefore, $b \lneqq a$ $(b < a)$.$)$

4) $\begin{cases} x < a \\ x > b \end{cases}$ 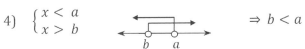 $\Rightarrow b < a$

$(\because$ If $b = a$, then $\begin{cases} x < a \\ x > a \end{cases}$ $\therefore$ There is no solution. Therefore, $b \lneqq a$ $(b < a)$.$)$

(7) For any $a > 0$,

$\begin{cases} |x| \le a \Rightarrow -a \le x \le a \\ |x| > a \Rightarrow x > a \text{ or } x < -a \end{cases}$

 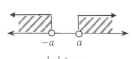

$|x| \le a$ $\qquad$ $|x| > a$

9-2 Graphs of Linear Inequalities with Two Variables

1. Half-planes

The graph of a linear inequality with two variables (x and y) consists of points in the coordinate plane whose coordinate (x, y) makes the linear inequality true.

For a linear inequality $ax + by \geq c$ or $ax + by \leq c$,

the coordinates on the line $ax + by = c$ are solutions to the linear inequality, so we use a straight line.

For a linear inequality $ax + by > c$ or $ax + by < c$,

the coordinates on the line $ax + by = c$ are not solutions to the linear inequality, so we use a dotted line instead of a straight line.

The straight line or dotted line separates the coordinate plane into two half-planes.

The inequality symbols "$\geq$" or "$\leq$" represent closed half-planes and the inequality symbols "$>$" or "$<$" represent open half-planes.

2. Graphing Linear Inequalities

Each half-plane region is the solution to the inequality.

To find the solution, choose a point in one half-plane and substitute it into the inequality.

If the point makes the inequality true, then the region including the point is the solution to the inequality. Shade the solution region.

Example

Graph the linear inequality $2x + y < 4$.

Consider the graph of the line $y = -2x + 4$.

The line $y = -2x + 4$ has slope -2 and y-intercept $(0, 4)$.

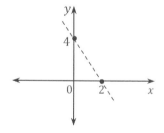

Choose a point $(-1, 0)$ in one half-plane and substitute the point into the inequality.

Since $2x + y = 2 \cdot (-1) + 0 = -2$ and $-2 < 4$, the inequality is true.

So the region including the point $(-1, 0)$ is the solution to $-2x + y < 4$.

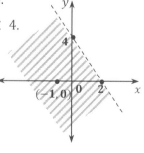

Example

Graph the linear inequality $2x - 3y \geq 6$.

$2x - 3y \geq 6 \quad \Rightarrow \quad 3y \leq 2x - 6 \Rightarrow \quad y \leq \frac{2}{3}x - 2$

Graph the line $y = \frac{2}{3}x - 2$.

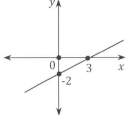

The line $y = \frac{2}{3}x - 2$ has x-intercept $(3, 0)$ and y-intercept $(0, -2)$.

Choose a point $(0, 0)$ as a testing point.

Substituting $x = 0$ and $y = 0$ into the inequality,

we get $2x - 3y = 2 \cdot 0 - 3 \cdot 0 = 0$.

Since $0 \geq 6$ is not true,

the region including the point $(0, 0)$ is not the solution.

Shade the region which does not include the point $(0, 0)$.

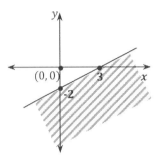

3. Graphing Systems of Inequalities

To solve a system of inequalities, find the solution region for each inequality.

Graph all inequalities on one coordinate plane. The overlapping region of the graph is the solution to the system of inequalities.

Every point in the overlap region makes all the inequalities true.

Example   Graph a system of linear inequalities $\begin{cases} 2x - y \geq 1 \\ x - 3y > 4 \end{cases}$.

Since $2x - y \geq 1 \quad \Rightarrow \quad y \leq 2x - 1$,

graph the line $y = 2x - 1$.

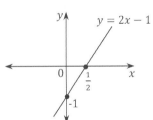

Choose a testing point $(0, 0)$.

Substituting $x = 0$ and $y = 0$ into the inequality,

$2x - y = 2 \cdot 0 - 0 = 0 \geq 1$: This is not true.

So,

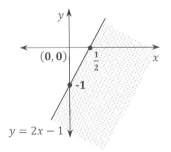

The shaded region is the solution to the linear inequality $2x - y \geq 1$.

Since $x - 3y > 4 \;\Rightarrow\; y < \frac{1}{3}x - \frac{4}{3}$, graph the line $y = \frac{1}{3}x - \frac{4}{3}$.

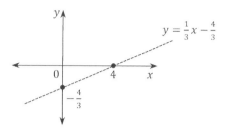

Choose a testing point $(0, 0)$.

Substituting $x = 0$ and $y = 0$ into the inequality,

$x - 3y = 0 - 3 \cdot 0 = 0 > 4$: This is not true.

So,

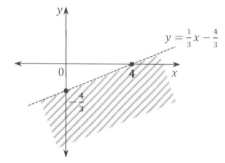

Graph both inequalities on the same coordinate plane. Shade the overlapping solution region more darkly than the other solution regions.

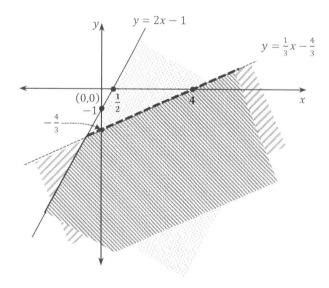

Exercises

#1. Find the range of x for the following systems:

(1) $\begin{cases} x \geq -2 \\ x < 4 \end{cases}$

(2) $\begin{cases} x > -3 \\ x > 0 \end{cases}$

(3) $\begin{cases} x < -1 \\ x \leq 2 \end{cases}$

(4) $\begin{cases} x \leq 1 \\ x \geq 3 \end{cases}$

(5) $\begin{cases} x \geq 2 \\ x < 2 \end{cases}$

(6) $\begin{cases} x \leq -3 \\ x \geq -3 \end{cases}$

#2. Solve the following systems:

(1) $\begin{cases} 3x - 1 < 4x + 2 \\ x + 3 \geq -2x \end{cases}$

(2) $\begin{cases} -2x + 4 > 3x - 6 \\ 3x - 5 \geq x + 3 \end{cases}$

(3) $\begin{cases} x - 2 > -5 \\ 5x - 3 < 2x + 3 \end{cases}$

(4) $\begin{cases} 4x - 3 > 5x - 2 \\ 2x - 2 > 3x - 5 \end{cases}$

(5) $\begin{cases} \frac{x+2}{2} \leq \frac{x-2}{3} + x \\ 2(x - 2) - \frac{x}{2} > 2x - 6 \end{cases}$

(6) $x - 3 < \frac{x}{2} + 1 \leq 3x - 1$

(7) $0.5x + \frac{x}{2} + 1 > 0.7x - 0.2 > 0.3x - 0.6$

(8) $\begin{cases} 2.1x > 3.6 - 0.9x \\ 3(x - 1) - 2 < \frac{1}{2}(3x + 1) \end{cases}$

#3. Find the range of $y = -2x + 3$ when x satisfies the following systems:

(1) $\begin{cases} x - 3 \leq 2 \\ -3x + 2 < 4 \end{cases}$

(2) $\begin{cases} 2x - 3 < 4x - 1 \\ 5x - 2 \geq 3x + 2 \end{cases}$

(3) $\frac{x-1}{3} \leq \frac{1}{4}(x - 3) < \frac{5-3x}{4}$

(4) $\begin{cases} 2.1x > 3.6 - 0.9x \\ 3(x - 1) - 2 < \frac{1}{2}(3x + 1) \end{cases}$

#4. Find the value of k for the following conditions:

(1) The system $\begin{cases} x + 5 < 2k \\ 3x - 2 \geq 4 \end{cases}$ has the solution $2 \leq x < 5$.

(2) The system $\begin{cases} \frac{2x+1}{3} > \frac{x-3}{5} \\ 0.6x - 2.4 < kx - 0.8 \end{cases}$ has the solution $-2 < x < 2$.

(3) The system $\begin{cases} -x + 2 \leq 0 \\ \frac{x}{2} + 3 \leq -k + 5 \end{cases}$ has the solution $x = 2$.

(4) The system $\begin{cases} \frac{k-x}{2} \leq x + 5 \\ 3 - 2x < 3x - 2 \end{cases}$ has the solution $x \geq 3$.

(5) The system $\begin{cases} 2x + 3 \le 4x - 5 \\ 3(x - k) \le x + 3 \end{cases}$ has only one solution.

(6) The system $3 < \dfrac{k-4x}{-2} < 5$ has the solution $1 < x < 2$.

#5. Find the range of k for the following conditions:

(1) The system $\begin{cases} 2x \le 5 - k \\ 3x - 3 \ge 2x - 1 \end{cases}$ has no solution.

(2) The system $\begin{cases} x - 3 \le 2x - 6 \\ 5x + k < 3x + 1 \end{cases}$ has no solution.

(3) The system $\begin{cases} 2x + 3 \le -5 \\ x + k > 1 \end{cases}$ has only one integer in the solution.

(4) The system $\begin{cases} x - 3 \ge 0 \\ 3x + k \le 2x + 3 \end{cases}$ has solutions.

#6. Find the sum of all integers that satisfy the following systems:

(1) $\begin{cases} 3x - 5 \le 7 \\ \frac{x-1}{2} < x + 3 \\ 2x - 5 < 5x + 4 \end{cases}$

(2) $\begin{cases} |x| \le 5 \\ |x| > 2 \end{cases}$

#7. The system $\begin{cases} x - 1 \ge 2x - 4 \\ \frac{x+k}{2} < 3x - 2 \end{cases}$ has 5 integers in the solution.

What is the minimum value for k?

#8. 3 more than twice a number is less than or equal to 8, and -1 is less than one fourth of the number.

(1) Find the range of the number.

(2) Find the sum of the maximum integer and minimum integer that satisfies the system.

(3) Solve the system with the condition, -1 is greater than one fourth of the number, instead of the second inequality in the system shown in (1). Find the maximum integer that satisfies the new system.

#9. The sum of three consecutive positive integers is greater than 60 and less than 65.

Find the largest number.

#10. The lengths of three sides of a triangle are $x - 4$, $x + 1$, and $x + 3$. Find the range of x.

#11. Nichole wants to produce new salt solution by adding salt into 20 ounces of a 15% salt solution. It will be a salt concentration greater than that of a 20% salt solution and less than that of a 25% salt solution. How much salt does she need to add?

#12. Richard jogs 12 miles. He begins by walking at a speed of 5 miles per hour. He then runs at a speed of 10 miles per hour for the remaining distance. If he wants to take at least 1 hour 45 minutes and at most 2 hours to complete his route, what is the longest distance he should walk?

#13. Nichole wants to reorganize all the books in her bookshelf. If she puts 30 books in each shelf, then 5 books will be left over. If she puts 35 books in each shelf, then there will be at least 20 books, but less than 25 books on the last shelf.   How many shelves are in the bookshelf?

#14. Solve the following inequalities:

(1) $|x - 3| < 4$

(2) $|x + 2| < 3x - 4$

(3) $|x + 2| + |3 - x| > 10$

#15 Graph the following systems of linear inequalities:

(1) $\begin{cases} y < x \\ y \geq -x \end{cases}$

(2) $\begin{cases} 2x + y \leq 4 \\ x - y \geq 2 \end{cases}$

(3) $\begin{cases} 3x + y \leq 6 \\ 2x + 3y > 4 \end{cases}$

(4) $\begin{cases} y \leq 2 \\ x + y > 3 \\ x > -4 \end{cases}$

Chapter 10. Linear Functions

10-1 Linear Functions and their Graphs

> A linear function is a function that can be graphically represented in a coordinate plane by a straight line.

1. Linear Functions

A function f is called a *linear function of* x if it is formed by

$f(x) = ax + b$ (Function form) or

$y = ax + b$ (Equation form), where $a(\neq 0)$, b are constants.

> Linear:
> The highest power in a polynomial is 1.

Note: $f(x) = ax + b$, $a \neq 0$ *is a first-degree polynomial function or linear function.*

$f(x) = ax^2 + bx + c$, $a \neq 0$ *is a second-degree polynomial function or quadratic function.*

For example, $y = 2x = 2x^1$, $y = \frac{1}{2}x + 1 = \frac{1}{2}x^1 + 1$: *Linear functions*

(Expressed in terms of x with power 1)

$y = \frac{2}{x} = 2x^{-1}$, $y = x^2 + 1$, $y = x^3 + 2$: *Not linear functions*

> For any constants, $a(\neq 0)$, b,
> $ax + b$: Expression
> $ax + b = 0$: Linear equation
> $ax + b > 0$: Linear inequality
> $y = ax + b$: Linear function

2. Graphing $y = ax + b$, $a \neq 0$

(1) Graphing $y = ax + b$, $a \neq 0$ using Translation

The graph of the function $y = ax + b$, $a \neq 0$

is a translation of the graph of the function $y = ax$

with the b units along the y-axis.

> The graph of the function f is the graph of the equation $y = f(x)$.

① $b > 0$

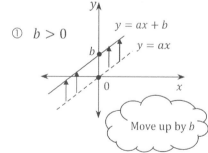

> Move up by b

② $b < 0$

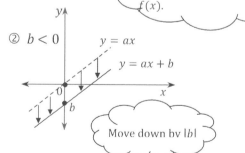

> Move down by $|b|$

(2) Intercepts

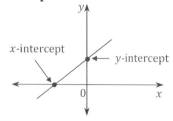

1) The *x-intercept* is the x-coordinate of the point where the line crosses (intersects) the x-axis. The *x*-intercept is the value of x when $y = 0$.

Example $y = 2x + 3 \xLongrightarrow{y=0} 0 = 2x + 3 \Rightarrow x = -\frac{3}{2}$

$\therefore x$-intercept is $-\frac{3}{2}$.

2) The *y-intercept* is the *y*-coordinate of the point where the line crosses (intersects) the *y*-axis. The *y*-intercept is the value of *y* when $x = 0$.

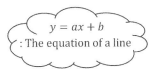

The x-intercept is on the x-axis and the y-intercept is on the y-axis.

Example $y = 2x + 3 \xLongrightarrow{x=0} y = 3$

$\therefore y$-intercept is 3.

(3) Graphing $y = ax + b$, $a \neq 0$ using Intercepts

$y = ax + b$: The equation of a line

Steps: 1) Identify the intercepts $\begin{bmatrix} x-\text{intercept: } 0=ax+b \; ; \; x=-\dfrac{b}{a} \\ y-\text{intercept: } y=a\cdot 0+b \; ; \; y=b \end{bmatrix}$

2) Graph the intercepts; Plot the points $\left(-\dfrac{b}{a}, 0\right)$ and $(0, b)$ on a coordinate plane.

3) Connect the points into a straight line.

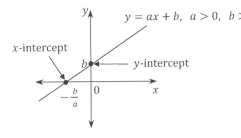

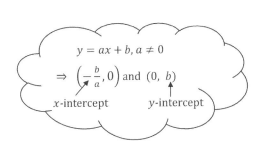

$y = ax + b, a \neq 0$

$\Rightarrow \left(-\dfrac{b}{a}, 0\right)$ and $(0, b)$

x-intercept $\quad y$-intercept

3. Horizontal and Vertical Lines

(1) A *horizontal line* passes through a point on the *y*-axis and is parallel to the *x*-axis.

For any constant k, $y = k$ intersects the *y*-axis at k and the value of x doesn't matter at all.

(2) A *vertical line* passes through a point on the *x*-axis and is parallel to the *y*-axis.

For any constant k, $x = k$ intersects the *x*-axis at k and the value of y doesn't matter at all.

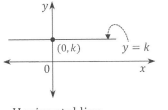

Horizontal line

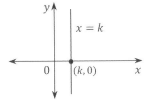

Vertical line

Note: *The line $y = 0$ is the x-axis and the line $x = 0$ is the y-axis.*

    A line passing through (k, p) and (k, q) is

    $x = k$ ($\because$ *The same x-coordinate*).

    A line passing through (p, k) and (q, k) is $y = k$ ($\because$ The same y-coordinate).

4. The Slope of a Line

To find an equation of a line, consider the slope of the line.

The *slope* of a line measures the steepness and direction of a line. The absolute value of the slope determines a line's steepness. The sign of the slope determines a line's direction.

(1) The Concept of Slope

A line which is not parallel to a coordinate axis may rise from lower left to upper right, or it may fall from upper left to lower right.

For a line passing through the given points (x_1, y_1) and (x_2, y_2), where $x_1 \neq x_2$, the positive number $y_2 - y_1$ is called the *rise*, and the positive number $x_2 - x_1$ is called the *run*. We define the *slope* (m) of the line by

$$\boxed{m = \frac{y_2 - y_1}{x_2 - x_1} \quad \text{or} \quad m = \frac{y_1 - y_2}{x_1 - x_2}}$$

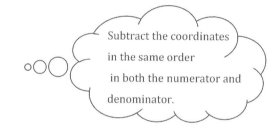

Subtract the coordinates in the same order in both the numerator and denominator.

Note : $Slope = \dfrac{rise}{run} = \dfrac{change\ in\ y}{change\ in\ x}$

         $= Ratio\ of\ the\ vertical\ change\ of\ a\ line\ and\ the\ horizontal\ change\ of\ a\ line.$

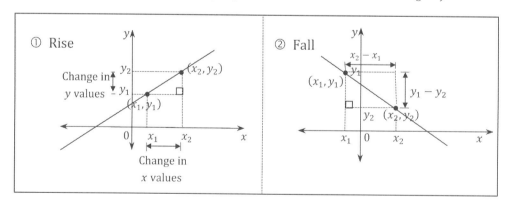

Note 1: *For any horizontal line $y = k$, the slope is 0.*

    ($\because$ *All the points (coordinates) on the line $y = k$ have same y value, k.*

    That is, there is no change in y.

    So, slope $= \dfrac{0}{Change\ in\ x} = 0$)

Note 2: For any vertical line $x = k$, the slope does not exist.

$(\because$ *All the points (coordinates) on the line $x = k$ have same x value, k.*

That is, there is no change in x.

So, slope $= \dfrac{change\ in\ y}{0}$.

This equation is undefined because a denominator can't be divided by zero.

Therefore, vertical lines have no slope (Slope is undefined).)

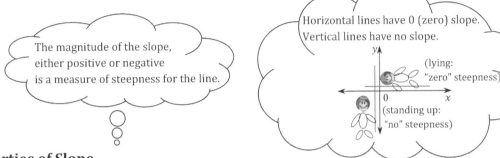

(2) Properties of Slope

1) For two lines $y = ax + b, a \neq 0$ and $y = cx + d, c \neq 0$,

the line $y = ax + b$ is steeper than the line $y = cx + d$ if $|a| > |c|$

The larger the absolute value of the slope, the steeper the line will be.

Example

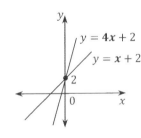

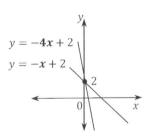

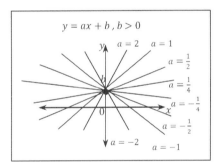

2) For a linear function $y = ax + b, a \neq 0$, the slope is a.

① $a > 0$ (Positive slope): The line goes uphill from left to right (the line rises to the right.)

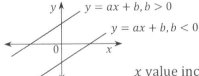

x value increases. $\Rightarrow$ y value increases.

② $a < 0$ (Negative slope): The line goes downhill from left to right (the line falls to the right.)

x value increases. $\Rightarrow$ y value decreases.

$y = ax + b,\ b > 0$

$y = ax + b,\ b < 0$

(3) Graphing $y = ax + b$, $a \neq 0$ using slope and y-intercept

1) Plot the y-intercept on a coordinate plane.

2) Using the slope, find the other point.

3) Connect the two points into a straight line.

Note: $y = ax + b$ $a > 0$, $b > 0$

① y-intercept b
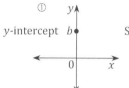

② Slope=a=$\frac{a}{1}$ $+1$
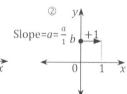

③ b $+a$ *move up*
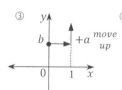

④ $a + b$ b
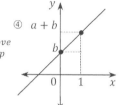

Example $y = -\dfrac{5}{2}x + 1$

① y-intercept: 1
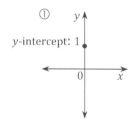

② Slope=$-\dfrac{5}{2}$ $+2$
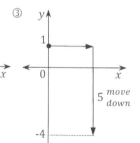

③ 5 *move down* -4
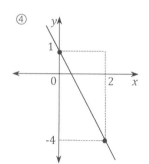

④ -4
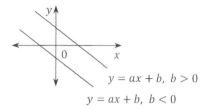

Slope > 0

move up
move right
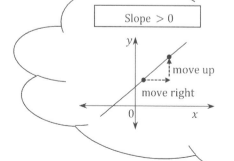

Slope < 0

move right
move down
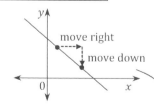

10-2 Lines and their Equations

For a given linear equation $ax + by + c = 0$ (constants $a \neq 0$, $b \neq 0$, and c),

the *linear function* is represented by

$$y = -\frac{a}{b}x - \frac{c}{b} \quad \text{(constants } a \neq 0, \ b \neq 0, \ \text{and } c \text{)}.$$

The graphs of the linear equation and the linear function are the same lines.

> To find a slope and y-intercept,
> transform the standard form, $ax + by + c = 0$
> into the form of $y = mx + P$.

Note 1: The standard form of an equation for a line is

$$ax + by + c = 0, \ a \neq 0 \ or \ b \neq 0.$$

However, $y = ax + b$ is not the standard form of a line.

$(\because$ *If a (x-coefficient$) = 0$, then $y = b$ (a horizontal line).*

However, vertical line $x = k$ can't be obtained

because the y-coefficient is $1(\neq 0)$, that is, the y- term does not disappear.)

Note 2: Graphing $ax + by + c = 0$:

① $a \neq 0, \ b \neq 0 \ \Rightarrow \ y = -\dfrac{a}{b}x - \dfrac{c}{b}$ *Line with slope , $-\dfrac{a}{b}$, and y-intercept, $-\dfrac{c}{b}$*

② $a \neq 0, \ b = 0 \ \Rightarrow \ ax + c = 0 \ ; \ x = -\dfrac{c}{a}$ *Vertical line parallel to the y-axis.*

③ $a = 0, \ b \neq 0 \ \Rightarrow \ by + c = 0 \ ; \ y = -\dfrac{c}{b}$ *Horizontal line parallel to the x-axis.*

1. Methods of Creating the Equations of Lines

*Note: **Intercepts***

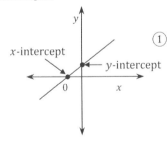

① x-intercept is the x-coordinate of the point where the line crosses (intersects) the x-axis.

x-intercept is the value of x when $y = 0$.

For example, $y = 2x + 3 \underset{y=0}{\Longrightarrow} 0 = 2x + 3 \ \Rightarrow \ x = -\dfrac{3}{2}$

$\therefore \ x$-intercept is $-\dfrac{3}{2}$.

② y-intercept is the y-coordinate of the point where the line crosses (intersects) the y-axis.

y-intercept is the value of y when $x = 0$.

For example, $y = 2x + 3 \underset{x=0}{\Longrightarrow} y = 3$

$\therefore \ y$-intercept is 3.

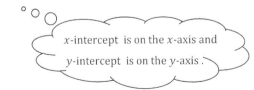

> x-intercept is on the x-axis and
> y-intercept is on the y-axis .

(1) A line which has slope m and y-intercept b (Slope-Intercept Form)

The equation of the line is $\boxed{y = mx + b}$

x-coefficient is slope m.

(2) A line which has slope m and passes through a point (x_1, y_1) (Point-Slope Form)

The equation of the line is $\boxed{y - y_1 = m(x - x_1)}$

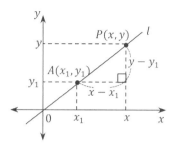

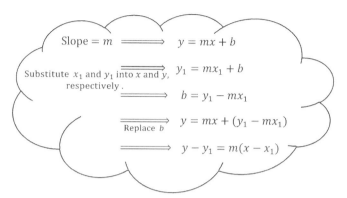

$$\text{Slope} = m \implies y = mx + b$$

$$\xRightarrow{\text{Substitute } x_1 \text{ and } y_1 \text{ into } x \text{ and } y,\ \text{respectively}.} y_1 = mx_1 + b$$

$$\implies b = y_1 - mx_1$$

$$\xRightarrow{\text{Replace } b} y = mx + (y_1 - mx_1)$$

$$\implies y - y_1 = m(x - x_1)$$

Let $P(x, y)$ be a point on a line l.

Then, the slope of $\overline{AP}$ is $m = \dfrac{y - y_1}{x - x_1}$.

Thus, the equation of the line l is $y - y_1 = m(x - x_1)$.

Therefore, any point $P(x, y)$ on the line l satisfies the equation.

Conversely, any point $P(x, y)$ that satisfies the equation is on the line l.

The equation of the line which passes through a point $A(x_1, y_1)$ and parallels to the y-axis

is $x = x_1$.

Hence, a line which has slope m and passes through a point $A(x_1, y_1)$ is $y - y_1 = m(x - x_1)$.

(3) A line which passes through two different points (x_1, y_1) and (x_2, y_2), $x_1 \neq x_2$

$\Rightarrow$ ① Find slope m

$$m = \frac{y_2 - y_1}{x_2 - x_1} \quad \text{or} \quad m = \frac{y_1 - y_2}{x_1 - x_2}$$

② Use the slope m and one of the two points. Using point-slope form,

$$y - y_1 = m(x - x_1) \quad \text{or} \quad y - y_2 = m(x - x_2)$$

Therefore, the equation of the line is given by

$$\boxed{\begin{array}{l} \text{When } x_1 \neq x_2, \ y - y_1 = \dfrac{y_2 - y_1}{x_2 - x_1}(x - x_1) \\[2mm] \text{When } x_1 = x_2, \ x = x_1 \end{array}}$$

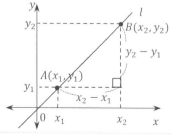

Case 1. When $x_1 \neq x_2$,

The slope of $\overline{AB}$ is $\dfrac{y_2 - y_1}{x_2 - x_1}$.

Since the line passes through a point $A(x_1, y_1)$,

The equation of the line is $y - y_1 = \dfrac{y_2 - y_1}{x_2 - x_1}(x - x_1)$.

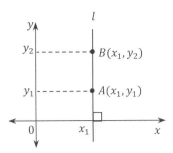

Case 2. When $x_1 = x_2$,

$\overline{AB}$ is parallel to the y-axis and

passes through a point $A(x_1, y_1)$.

The equation of the line is $x = x_1$.

(4) A line which has an x-intercept p and y-intercept q

The same as a line which passes through two points $(p, 0)$ and $(0, q)$

$\Rightarrow$ ① Find slope m

$$m = \frac{q-0}{0-p} \ \text{ or } \ m = \frac{0-q}{p-0}$$

② Use the line which has slope m and y-intercept q:

$$y = mx + q$$

Therefore, the equation of the line is given by

$$\boxed{\frac{x}{p} + \frac{y}{q} = 1 \ \ (pq \neq 0)}$$

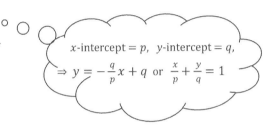

x-intercept $= p$, y-intercept $= q$,

$\Rightarrow y = -\frac{q}{p}x + q$ or $\frac{x}{p} + \frac{y}{q} = 1$

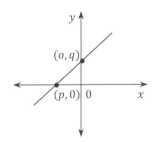

Using the equation of the line which has slope m

and y-intercept q, we have

$$y = mx + q = -\frac{q}{p}x + q$$

$$\therefore \ \frac{q}{p}x + y = q$$

$$\therefore \ \frac{x}{p} + \frac{y}{q} = 1 \ \ (pq \neq 0)$$

Example   Two lines $\frac{x}{3} + \frac{y}{4} = 1$ and $y = mx$ intersect at a point P.

When the ratio of the areas of $\triangle OAP$ and $\triangle OBP$ is $2 : 3$, find the value of m.

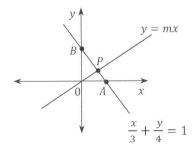

$\because$ The x-axis and y-axis of the line $\frac{x}{3} + \frac{y}{4} = 1$

are 3 and 4, respectively.

$\therefore \ A = A(3, 0)$ and $B = B(0, 4)$

Since (The area of $\triangle OAP$) : (The area of $\triangle OBP$) = 2 : 3,

the point P divides the segment $\overline{AB}$ internally in the ratio 2 : 3 (i.e., $\overline{AP} : \overline{PB} = 2 : 3$)

∴ The coordinates of point P are

$$x = \frac{2\cdot 0 + 3\cdot 3}{2+3} = \frac{9}{5}, \quad y = \frac{2\cdot 4 + 3\cdot 0}{2+3} = \frac{8}{5}$$

Since the slope of the line which passes through the origin $(0, 0)$ and $\left(\frac{9}{5}, \frac{8}{5}\right)$ is

$\frac{\frac{8}{5}-0}{\frac{9}{5}-0} = \frac{8}{9}$, we have the slope $m = \frac{8}{9}$.

> The line $y = 0$ is the x-axis and the line $x = 0$ is the y-axis.

(5) Lines and Their Equations

The two equations of lines $y - y_1 = m(x - x_1)$ and $x = x_1$ are represented by

$mx - y - mx_1 + y_1 = 0$ and $x - x_1 = 0$, respectively.

Therefore, an equation of a line is expressed as a linear equation for x and y:

$ax + by + c = 0$ $(a \neq 0$ or $b \neq 0)$

Conversely, a line equation for x and y : $ax + by + c = 0$ $(a \neq 0$ or $b \neq 0)$ is expressed as

equations of lines: i) When $b \neq 0$, $y = -\frac{a}{b}x - \frac{c}{b}$

ii) When $b = 0$, $x = -\frac{c}{a}$

Equations of Lines

❶ Vertical line $x = k$

❷ Horizontal line $y = k$

❸ General linear equation $ax + by + c = 0$

❹ Slope-Intercept form $y = mx + b$

❺ Point-Slope form $y - y_1 = m(x - x_1)$

> Parallel lines (⟋⟋) never intersect even when extended. Perpendicular lines (⟷) intersect at one point and form a right angle.

2. Parallel and Perpendicular Lines

For any two different lines $y = m_1 x + a$ and $y = m_2 x + b$, parallel and perpendicular lines are identified by comparing their slopes, m_1 and m_2.

(1) $m_1 = m_2$ **The slopes are the same.**

⇒ **Parallel lines**

(2) $m_1 \cdot m_2 = -1 \left(\text{or } m_1 = -\frac{1}{m_2}; \ m_2 = -\frac{1}{m_1}\right)$ **Slopes are negative reciprocals of each**

other. ⇒ **Perpendicular lines**

Note 1:   For a line i which has the equation $y = \frac{2}{3}x + 2$, the slope of the line i is $\frac{2}{3}$.

So, the parallel line j has the same slope of $\frac{2}{3}$, and the slope of a perpendicular line k is $-\frac{3}{2}$.

Note 2 :   ① Lines i and j are parallel.

⇒ These two lines have the same slope and different y-intercepts.

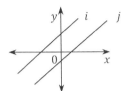

The lines never intersect

② Lines i and k are perpendicular.

⇒ Their slopes are negative reciprocals of each other.

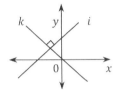

The lines intersect at one point and form a right angle (90°).

③ Lines i and l coincide.

⇒ These two lines have the same slope and the same y-intercept.

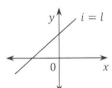

Note 3 :   ① These lines are symmetrical with respect to the x-axis.

This happens whenever (x, y) and $(x, -y)$ are on the same graph.

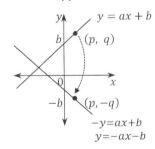

② These lines are symmetrical with respect to the y-axis.

This happens whenever (x, y) and $(-x, y)$ are on the same graph.

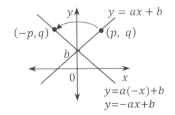

3. Translation and Symmetry for $y = ax$, $a \neq 0$

(1) $(x, y) \longrightarrow (x + m,\ y + n)$: Translate m units along the x-axis and n units along the y-axis

$$\Rightarrow y - n = a(x - m)$$

(2) Symmetry

1) along the x-axis $(x, y) \longrightarrow (x, -y)$

2) along the y-axis $(x, y) \longrightarrow (-x, y)$

3) along the origin $(0, 0)$ $(x, y) \longrightarrow (-x, -y)$

4) When $y = x$ $(x, y) \longrightarrow (y, x)$

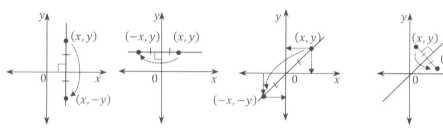

Figure 1)   Figure 2)   Figure 3)   Figure 4)

10-3 Solving Equations by Graphing

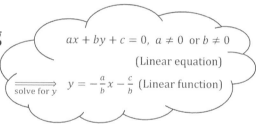

$$ax + by + c = 0,\ a \neq 0 \text{ or } b \neq 0$$
(Linear equation)

$\xrightarrow[\text{solve for } y]{}$ $y = -\dfrac{a}{b}x - \dfrac{c}{b}$ (Linear function)

For any constants a, b, and c,

$$ax + by + c = 0,\ a \neq 0 \text{ or } b \neq 0$$

is the standard form of equation for a straight line.

The graph of an equation with variables x and y consists of points in the coordinate plane whose coordinates (x, y) make the equation true.

The relationship of the two linear equations $\begin{cases} ax + by + c = 0 \\ a'x + b'y + c' = 0 \end{cases}$ in a coordinate plane is one of the following cases:

(1) Case 1: Parallel

The corresponding system of equations does not have a solution. The system is called *inconsistent*.

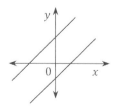

$\Rightarrow$ ① no intersection

② no solution for the system

③ same slopes $\left(-\dfrac{a}{b} = -\dfrac{a'}{b'}\ ;\ \dfrac{a}{b} = \dfrac{a'}{b'}\ ;\ ab' = a'b\ ;\ \dfrac{a}{a'} = \dfrac{b}{b'}\right)$

④ different y-intercepts $\left(-\dfrac{c}{b} \neq -\dfrac{c'}{b'}\ ;\ \dfrac{c}{b} \neq \dfrac{c'}{b'}\ ;\ b'c \neq bc'\ ;\ \dfrac{b}{b'} \neq \dfrac{c}{c'}\right)$

(2) Case 2: Intersecting at one point (perpendicular or not)

The corresponding system of equations has exactly one solution. The system is called *consistent and independent*, shortly *independent*.

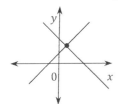

$\Rightarrow$ ① 1 intersection

② 1 solution for the system

③ different slopes $\left(-\dfrac{a}{b} \neq -\dfrac{a'}{b'}\right)$

(3) Case 3: Coinciding

The corresponding system of equations has an infinite number of solution. The system is called *consistent and dependent*, shortly *dependent*.

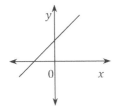

$\Rightarrow$ ① unlimited number of intersections

② unlimited number of solutions for the system

③ the same slopes $\left(-\dfrac{a}{b} = -\dfrac{a'}{b'}\right)$

④ the same y-intercepts $\left(-\dfrac{c}{b} = -\dfrac{c'}{b'}\right)$

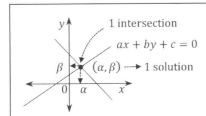

The solution $(x = \alpha, y = \beta)$ of two equations (a system of equations) with two variables is the intersection point (α, β) between the two lines.

Note: $\begin{cases} ax + by + c = 0 \\ a'x + b'y + c' = 0 \end{cases} \xrightarrow[\text{solve for } y]{} \begin{cases} y = mx + p \\ y = m'x + p' \end{cases}$

① $\dfrac{a}{a'} = \dfrac{b}{b'} \neq \dfrac{c}{c'}$ $\Rightarrow$ *parallel* $\Leftarrow$ $m = m'$ and $p \neq p'$

② $\dfrac{a}{a'} \neq \dfrac{b}{b'}$ $\Rightarrow$ *intersecting at one point* $\Leftarrow$ $m \neq m'$

③ $\dfrac{a}{a'} = \dfrac{b}{b'} = \dfrac{c}{c'}$ $\Rightarrow$ *coinciding* $\Leftarrow$ $m = m'$ and $p = p'$

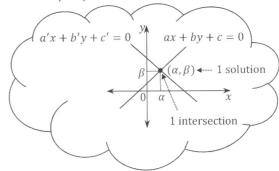

10-4 Graphing with Absolute Values

Note : $|a| = \begin{cases} a , & a \geq 0 \\ -a , & a < 0 \end{cases}$

For $y = |x|$, all real numbers have exactly one absolute value.

So, $y = |x|$ is a function.

1. Graphing of an absolute value function, $y = |x|$.

Case 1: $x \geq 0 \Rightarrow y = x$	Case 2: $x < 0 \Rightarrow y = -x$
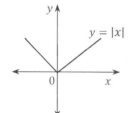	

Therefore, the graph of $y = |x|$ is

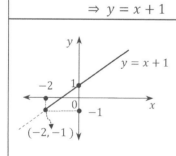

where, domain = all real numbers and

range = $\{ y| y \geq 0 \}$; all nonnegative real numbers.

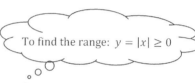

2. Graphing $y = |x + 2| - 1$

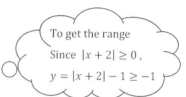

Case 1 : $x + 2 \geq 0 \ (x \geq -2)$ $\Rightarrow y = x + 1$	Case 2 : $x + 2 < 0 \ (x < -2)$ $\Rightarrow y = -x - 3$
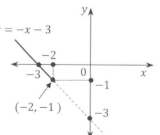	

Therefore, the graph of $y = |x + 2| - 1$ is

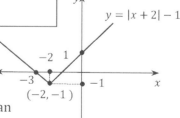

where, domain = all real numbers and

range = $\{ y| y \geq -1 \}$: all real numbers greater than or equal to -1.

Exercises

#1 Identify linear functions. Mark O for a linear function or × for a non-linear function.

(1) $y = 2 + x$

(2) $2x + y = 3$

(3) $y = \dfrac{2}{x} + 1$

(4) $x^2 + y^2 = 1$

(5) $y = x^2 + 2x + 1$

(6) $x + y = 0$

(7) $y = 2(x + 1) - 2x$

(8) $y = x^2 - (x + 2)^2$

(9) $y = 1$

(10) $xy = 2$

#2 Find the following values for the linear function $f(x) = 2x + 3$

(1) $f(0)$

(2) $f(1) + f(-1)$

(3) $\dfrac{1}{2}f(-2) \cdot f\left(\dfrac{1}{2}\right)$

(4) $f(f(2))$

#3 Find the following values for the given linear functions with a condition

(1) $f(1)$ for $f(x) = 2ax + 1$ with $f(-1) = 3$

(2) $\dfrac{a}{2}$ for $f(x) = \dfrac{1}{2}x + 5$ with $f\left(\dfrac{a}{2}\right) = -a$

(3) $a + b$ for $f(x) = 3ax - 2$ with $f(-1) = 4$ and $f(b) = 1$

(4) $a + \dfrac{1}{a}$ when $f(x) = 3x - 1$ passes through the point $(a, a + 3)$.

(5) $a - b$ when $f(x) = ax + 2$ passes through both point $(1, 3)$ and point $(2, b)$.

#4 Find the value of $a + b$ for which

(1) The graph of $y = ax + 2$ is translated by b along the y-axis from a graph of $y = 3x - 5$.

(2) The graph is translated by a along the y-axis from a graph of $y = 2x + 4$ and passes

through both point $(a + 1, -2)$ and point $\left(-\dfrac{1}{3}, b\right)$.

(3) A point $(-1, 1)$ is on the graph of $y = -2x + a$. If the graph is translated by b along the y-axis, then it will pass through the point $(3, -4)$.

#5 Find the x-intercept and y-intercept.

(1) The linear function $y = ax + b$ passes through both point $(1, 2)$ and point $(-1, 4)$.

(2) The graph of $y = ax + b$ intersects the graph of $y = 2x + 3$ on the x-axis. It also intersects the graph of $y = -5x - 6$ on the y-axis.

(3) The area surrounded by the graph of $y = \frac{1}{2}x + a$ $(a > 0)$, the x-axis, and the y-axis is 36.

#6 Find the area of the polygon surrounded by

(1) A graph of $y = -\frac{2}{3}x + 2$, the x-axis, and the y-axis

(2) The graphs of $y = x + 4$ and $y = -\frac{1}{2}x + 4$ and the x-axis

(3) The graphs of $y = \frac{1}{3}x + 3$ and $y = \frac{1}{9}x + 1$ and the y-axis

(4) The graphs of $x = 3$ and $y = 4$, the x-axis, and the y-axis

#7 Find the slopes of the lines containing the given two points.

(1) $(2, 3)$ and $(4, -1)$

(2) $(0, 5)$ and $(-5, 0)$

(3) $(4, -1)$ and $(2, -5)$

(4) $(-3, 2)$ and $(0, -1)$

(5) $(-2a, 0)$ and $(0, -4a), a \neq 0$

#8 Find the slope m and y-intercept b of each line.

(1) $2x + 3y = 4$

(2) $3y = 4x - 5$

(3) $x + 3y = -2$

(4) $y + 5 = 3x$

(5) $y = 2x$

(6) $y - 2 = 0$

#9 Find an equation in the standard form for each line.

(1) with y-intercept -3 and slope 2

(2) with y-intercept 5 and slope 0

(3) with x-intercept 5 and slope $-\frac{2}{3}$

(4) with x-intercept -3 and slope -2

(5) through $(1, 2)$ with slope 3

(6) through $(3, -4)$ with slope -2

(7) through $(2,3)$ with undefined slope

(8) through $(-2,3)$ with y-intercept -1

(9) through $(2,4)$ with x-intercept -5

(10) through $(3,1)$ and $(-2,4)$

(11) through $(-2,-3)$ and $(-1,5)$

(12) with x-intercept -3 and y-intercept 3

(13) with x-intercept $\frac{3}{2}$ and y-intercept -4

(14) Vertical line through $(-1,2)$

(15) Horizontal line through $(3,-4)$

#10 Find an equation for the line through $(2,3)$ which is

(1) parallel to the line $y = 2x - 5$

(2) parallel to the line $y = -3x + 1$

(3) parallel to the line $x = 4$

(4) parallel to the line $y = -2$

(5) parallel to the line $3x + 4y = 5$

(6) perpendicular to the line $y = \frac{2}{3}x - 1$

(7) perpendicular to the line $x + 3y = -3$

(8) perpendicular to the line $x = 5$

(9) perpendicular to the line $y = -2$

#11 Find the value of a for the following lines

(1) through $(2,3)$ and $(1,-a)$ with slope 2

(2) through $(2a - 1, -2)$ and $(-1,1)$ with slope -2

(3) through $(1,-2)$, $(-3,2)$, and $(-a + 1, -5)$

(4) through $(2a + 1, -4)$, $(2,5)$, and $(2,-3)$

(5) through $(-3,3)$, $(3, a - 1)$, and $(0,3)$

(6) through $(a, 2a - 3)$ and $(-a - 1, 3 + 4a)$ and parallel to x-axis

(7) through $(-3a + 1, -5)$ and $(2a - 1, a + 3)$ and perpendicular to x-axis

(8) through $(3, -2a)$ and $(2a - 1, -3a + 2)$ and parallel to the y-axis

(9) through $(-1,5)$ and $(2,-4)$ and parallel to the line $ax + 3y + 5 = 0$

#12 Find the value of a such that the line $ax + 2y = 5$

    (1) is parallel to the line $2x + 3y = -2$.

    (2) is perpendicular to the line $y = -2x + 3$.

    (3) coincides with the line $6y = -4x + 15$.

#13 Find the value of ab for which

    (1) the system $\begin{cases} x - 3y = a \\ 2x + by = 3 \end{cases}$ has the intersection point $(2, 3)$.

    (2) the system $\begin{cases} -ax + by = 4 \\ 2ax + 3by = 2 \end{cases}$ has the intersection point $(-1, 2)$.

    (3) the system $\begin{cases} px + y = 3 \\ 2x - 3y = q \end{cases}$ has no intersection when $p = a$, $q \neq b$.

    (4) the system $\begin{cases} 2ax + 4y = -3 \\ 3x + 6y = 2b \end{cases}$ has an unlimited number of intersections.

#14 Find the value of a such that

    (1) the system $\begin{cases} ax + y = -2 \\ -3x + 2y = 4 \end{cases}$ has no solution.

    (2) the system $\begin{cases} 2x - ay + 3 = 0 \\ x + 3y - 2 = 0 \\ 2x + y + 1 = 0 \end{cases}$ has one solution.

    (3) the system $\begin{cases} x - 3y = 2 \\ 2x + y = -3 \end{cases}$ has a solution $(2a, -1)$.

    (4) the line $2ax + 3y - 1 = 0$ passes through the intersection of the system $\begin{cases} x - 2y = 3 \\ 2x + 2y = 1 \end{cases}$.

#15 Find the equation of each line such that

    (1) the line passes through the intersection of the system $\begin{cases} x + 2y = 3 \\ 3x + y = -2 \end{cases}$

    and runs parallel to the y-axis.

    (2) the line passes through the intersection of the system $\begin{cases} -x + y + 2 = 0 \\ 2x + y - 3 = 0 \end{cases}$

    and runs perpendicular to the x-axis.

    (3) the line passes through the intersection of the system $\begin{cases} 2x - y + 3 = 0 \\ x + 2y + 4 = 0 \end{cases}$

    and runs parallel to the line $3x + 2y = 5$.

#16 Find the area of the polygon surrounded by the two lines ($3x + 4y - 16 = 0$ and $3x - 2y - 10 = 0$), the x-axis, and the y-axis.

#17 The area of a polygon surrounded by $y = x$, $y = ax + b$ ($b > 0$) which has x-intercept 6, and the x-axis is 12. Find the area of a polygon surrounded by those two lines and the y-axis.

#18 The perimeter of a rectangle with the length 5 inches and the width x inches is y square inches. Find the relationship between x and y.

#19 Richard drives a car 15 miles total, from place A to place B. He begins at a speed of 30 miles per hour. x minutes after departing, he has y miles more to go to arrive at place B. Find the relationship between x and y.

#20 Richard and Nichole drive toward each other from opposite starting points 4 miles apart. Richard drives at a speed of 40 miles per hour and Nichole drives at a speed of 35 miles per hour. After x minutes, the distance between the two is y miles. Find the relationship between x and y.

#21 Graph the following lines

(1) $y = -|x|$

(2) $y = -|x + 3| + 2$

(3) $y = |x| + x$

(4) $|y - 1| = x + 2$

Chapter 11. The Real Number System

11-1 Exponents and Radicals

1. Positive Integral Exponents

(1) Definition

For any real number a and positive integer n, the nth *power of* a is written as:

$$a^n = \underbrace{a \times a \times \cdots\cdots \times a}_{n \text{ factors}}$$ Product of n repeated factors of a.

In the expression a^n, n is called the *exponent* (or *power*) of a and a is called the *base*.

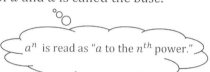

a^n is read as "a to the nth power."

(2) Properties of Exponents

For any real numbers $a, b,$ and positive integers $m, n,$

1) $a^m \times a^n = a^{m+n}$

2) $(a^m)^n = a^{mn}$

3) $(ab)^m = a^m b^m$

4) $\dfrac{a^m}{a^n} = a^{m-n}$

2. Negative Integral and Zero Exponents

(1) Definition

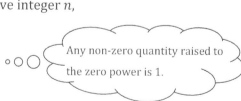

If n is positive, $-n$ is negative.

Consider $\dfrac{a^2}{a^5} = a^{2-5} = a^{-3}$

Since $\dfrac{a^2}{a^5} = \dfrac{1}{a^3}$, $\quad a^{-3} = \dfrac{1}{a^3}$

In general, for any real number a $(a \neq 0)$ and positive integer n,

$$a^{-n} = \dfrac{1}{a^n} \; ; \; a^0 = 1$$

For any non-zero number a, a^0 is equal to 1.

$(\because$ Since $\dfrac{a^2}{a^2} = a^{2-2} = a^0$ and $\dfrac{a^2}{a^2} = 1$, $a^0 = 1$)

For example, $2^0 = 1$ and $(-2)^0 = 1$

Any non-zero quantity raised to the zero power is 1.

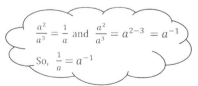

$\dfrac{a^2}{a^3} = \dfrac{1}{a}$ and $\dfrac{a^2}{a^3} = a^{2-3} = a^{-1}$

So, $\dfrac{1}{a} = a^{-1}$

(2) Properties of Exponents

For any real numbers $a(\neq 0), b(\neq 0)$, and positive integers m, n,

1) $a^m \times a^n = a^{m+n}$

2) $(a^m)^n = a^{mn}$

3) $(ab)^m = a^m b^m$

11-2 Radicals

1. Square Roots

(1) Definition

1) *A perfect square* is an integer that is equal to the product of which two identical numbers.

Example

$$2 \times 2 = 4$$
Perfect square

Two identical numbers

$$3 \times 3 = 9$$
Perfect square

Two identical numbers

2) The *square root* of a perfect square is equal to the number which is multiplied twice to get the perfect square and is represented by a radical sign " $\sqrt{}$ ".

Note : $\sqrt{a}$ means the positive square root of a number, a.

$-\sqrt{a}$ means the negative square root of a number, a.

Example

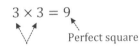

$\sqrt{4} = 2$

square root of 4

the number that is multiplied twice

$\sqrt{9} = 3$

square root of 9

the number that is multiplied twice

Note: ① Since $3 \times 3 = 9$ and $(-3) \times (-3) = 9$, square roots of 9 are 3 and -3.

② Since $\sqrt{a}$ represents the positive square root of a number, a, $\sqrt{9} = 3$.

But $\sqrt{9} = -3$ is incorrect.

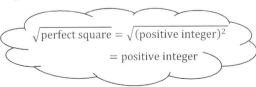

$\sqrt{\text{perfect square}} = \sqrt{(\text{positive integer})^2}$

$= \text{positive integer}$

3) In a symbol $\sqrt[n]{a}$, $\sqrt[n]{a}$ is called a *radical*, n is called an *index*, and a is called a *radicand*.

The index 2 of the square root is usually omitted. That is, $\sqrt[2]{a} = \sqrt{a}$.

Example

$5^2 = 5 \cdot 5 = 25$

Since 5 is one of the two equal factors of 25, 5 is a square root of 25.

Since $(-2)^3 = (-2) \cdot (-2) \cdot (-2) = -8$, -2 is a cube root of -8.

Since $3^4 = 3 \cdot 3 \cdot 3 \cdot 3 = 81$, 3 is a fourth root of 81.

4) For any number a,

$$
\boxed{
\begin{array}{l}
① \ \ a \geq 0 \ \Rightarrow \begin{cases} (\sqrt{a})^2 = a \ ; \ \ (-\sqrt{a})^2 = (\sqrt{a})^2 = a \\ \sqrt{a^2} = a \ \ ; \ \ \ \sqrt{(-a)^2} = \sqrt{a^2} = a \end{cases} \\[4mm]
② \ \ a < 0 \ \Rightarrow \sqrt{a^2} = -a \ \ (\text{positive integer})
\end{array}
}
$$

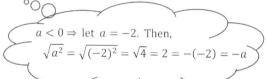

$a < 0 \Rightarrow$ let $a = -2$. Then,
$\sqrt{a^2} = \sqrt{(-2)^2} = \sqrt{4} = 2 = -(-2) = -a$

Note: For any number a,

$\sqrt{a^2} = |a| = \begin{cases} a, \ a \geq 0 \\ -a, \ a < 0 \end{cases}$, *where $|a|$ denotes the absolute value of a.*

$a^2 = b \Rightarrow a = \pm\sqrt{b} \ (+\sqrt{b} \ \text{or} - \sqrt{b})$

EX. $\ \ 3^2 = 9 \Rightarrow \ \ 3 = +\sqrt{9}$

$(-3)^2 = 9 \Rightarrow -3 = -\sqrt{9}$

$\sqrt{a^2}$ is always a positive number, whether a is positive or negative.

Example

$a = 3 \Rightarrow \sqrt{a^2} = \sqrt{3^2} = \sqrt{9} = 3 = a$ (Positive number)

$a = -3 \Rightarrow \sqrt{a^2} = \sqrt{(-3)^2} = \sqrt{9} = 3 = -(-3) = -a$ (Positive number)

Therefore, $\sqrt{a^2} = |a| \geq 0$

(2) Magnitude of Square Roots

For any positive numbers a and b,

$$
\boxed{
\begin{array}{l}
1) \ \ a < b \ \ \Rightarrow \ \ \sqrt{a} < \sqrt{b} \\[2mm]
2) \ \ \sqrt{a} < \sqrt{b} \ \ \Rightarrow \ \ a < b \\[2mm]
3) \ \ \sqrt{a} < \sqrt{b} \ \ \Rightarrow \ \ -\sqrt{b} < -\sqrt{a}
\end{array}
}
$$

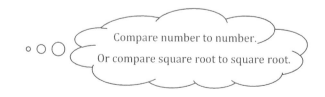

Compare number to number.
Or compare square root to square root.

Note : ① $0 < a < \sqrt{b} < c \Rightarrow a^2 < b < c^2$

② To compare a and $\sqrt{b}$, remove the radical sign squaring both comparing numbers.

That is, compare $(a)^2 = a^2$ and $(\sqrt{b})^2 = b$

Or convert a number into a radical. That is, compare $a = \sqrt{a^2}$ and $\sqrt{b}$.

Example   Compare 3 to $\sqrt{5}$.

Since $3^2 = 9$ and $(\sqrt{5})^2 = 5$, $3^2 > (\sqrt{5})^2$ $\therefore 3 > \sqrt{5}$

OR since $3 = \sqrt{9}$ and $9 > 5$, $\sqrt{9} > \sqrt{5}$ $\therefore 3 > \sqrt{5}$

2. Properties of Radicals (Roots)

$$1^2(=1) < \left(\sqrt{2}\right)^2(=2) < 2^2(=4) \Rightarrow 1 < \sqrt{2} < 2$$

$$2^2(=4) < \left(\sqrt{5}\right)^2(=5) < 3^2(=9) \Rightarrow 2 < \sqrt{5} < 3$$

Note   ① $\sqrt[n]{a} = a^{\frac{1}{n}}$: the nth root of a.

② The square root of a $(\sqrt{a})$: $\sqrt{a} = \sqrt[2]{a} = a^{\frac{1}{2}}$ (The index 2 is usually omitted.)

③ The cube root of a: $\sqrt[3]{a} = a^{\frac{1}{3}}$

For example, $\sqrt{4} = \sqrt[2]{4} = \sqrt[2]{2^2} = (2^2)^{\frac{1}{2}} = 2^1 = 2$,

$\sqrt[3]{8} = \sqrt[3]{2^3} = (2^3)^{\frac{1}{3}} = 2^1 = 2$, and $\sqrt[4]{16} = \sqrt[4]{2^4} = (2^4)^{\frac{1}{4}} = 2^1 = 2$

(1) For any non-negative real numbers a and b,

1) $\sqrt{a^2} = a$

2) $\sqrt{ab} = \sqrt{a}\sqrt{b}$

3) $\sqrt{\dfrac{a}{b}} = \dfrac{\sqrt{a}}{\sqrt{b}}$, $b \neq 0$

4) $\sqrt[n]{a} = a^{\frac{1}{n}}$

5) $\sqrt[n]{a^m} = (a^m)^{\frac{1}{n}} = a^{\frac{m}{n}}$

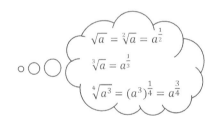

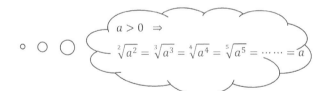

(2) For any real numbers a and b,

For a positive real number a, consider $\sqrt[3]{-a}$.

Since there is a negative real number $-b$ $(b > 0)$ such that $(-b)^3 = -a$,

$-b = \sqrt[3]{-a} = (-a)^{\frac{1}{3}}$.

1) $\sqrt[n]{a^n} = \begin{cases} |a|, & \text{if } n \text{ is even} \\ a, & \text{if } n \text{ is odd} \end{cases}$

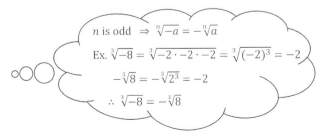

Example $a = -4 \Rightarrow$ $\begin{cases} \sqrt[2]{a^2} = \sqrt[2]{(-4)^2} = \sqrt[2]{16} = \sqrt[2]{4^2} = (4^2)^{\frac{1}{2}} = 4^1 = 4 = |-4| = |a| \\ \sqrt[3]{a^3} = \sqrt[3]{(-4)^3} = (-4)^{3 \cdot \frac{1}{3}} = (-4)^1 = -4 = a \end{cases}$

2) $\boxed{\sqrt{a^2} = |a|}$

3) $\boxed{\sqrt{a^3} = a\sqrt{a}}$

Example $\sqrt{2^3} = \sqrt{8} = \sqrt{2 \cdot 4} = \sqrt{2}\sqrt{4} = \sqrt{2} \cdot 2 = 2\sqrt{2}$

3. Solving Radical Equations

To solve a radical equation, eliminate the radical from the equation by raising an nth root to the n power.

Since the resulting equation is not equivalent to the given equation, you must check all solutions in the given equation.

(1) Solve the equation $\sqrt{x} = a$ for x.

If the radical is already isolated on one side of the equation, square both sides of the equation to eliminate the square root.

That is, $\sqrt{x} = a \Rightarrow (\sqrt{x})^2 = a^2 \Rightarrow x = a^2$

Now, check the solution in the original equation. That is, $\sqrt{x} = \sqrt{a^2} = a$.

So, $x = a^2$ makes the given equation, $\sqrt{x} = a$, true.

Therefore, $x = a^2$ is the solution for the given equation.

(2) Solve the equation $\sqrt{x} + a = b$ for x.

First, isolate the radical on one side and all the other terms on the other side of the equation. Then, square both sides to remove the radical.

That is, $\sqrt{x} + a = b \Rightarrow \sqrt{x} = b - a \Rightarrow (\sqrt{x})^2 = (b-a)^2 \Rightarrow x = (b-a)^2$

Now, check the solution in the original equation.

That is, $\sqrt{x} + a = \sqrt{(b-a)^2} + a = b - a + a = b$

So, $x = (b-a)^2$ makes the given equation, $\sqrt{x} + a = b$, true.

Therefore, $x = (b-a)^2$ is the solution for the given equation.

(3) If there are two radicals, separate the radical expressions.

You may need to square the expression twice to solve.

Example Solve the equation: $\sqrt{x+2} - \sqrt{2x+3} = 0$ for x.

$$\sqrt{x+2} - \sqrt{2x+3} = 0 \xRightarrow{\text{separating}} \sqrt{x+2} = \sqrt{2x+3} \xRightarrow{\text{squaring}} \left(\sqrt{x+2}\right)^2 = \left(\sqrt{2x+3}\right)^2$$

$$\Rightarrow \ x + 2 = 2x + 3 \ \therefore \ x = -1$$

$$\xRightarrow{\text{checking}} \sqrt{x+2} - \sqrt{2x+3} = \sqrt{-1+2} - \sqrt{-2+3} = \sqrt{1} - \sqrt{1} = 0$$

$$\therefore \ \sqrt{x+2} - \sqrt{2x+3} = 0 \text{ is true.}$$

Therefore, $x = -1$ is the solution for the given equation.

Example Solve the equation: $\sqrt{x+13} - \sqrt{7-x} = 2$ for x.

$$\sqrt{x+13} - \sqrt{7-x} = 2 \xRightarrow{\text{separating}} \sqrt{x+13} = 2 + \sqrt{7-x}$$

$$\xRightarrow{\text{squaring}} \left(\sqrt{x+13}\right)^2 = \left(2 + \sqrt{7-x}\right)^2$$

$$\Rightarrow \ x + 13 = 4 + 4\sqrt{7-x} + (7-x) \ \Rightarrow \ 2x + 2 = 4\sqrt{7-x}$$

$$\xRightarrow{\text{isolating}} 2\sqrt{7-x} = x + 1$$

$$\xRightarrow{\text{squaring}} \left(2\sqrt{7-x}\right)^2 = (x+1)^2 \ \Rightarrow \ 4(7-x) = x^2 + 2x + 1$$

$$\Rightarrow \ x^2 + 6x - 27 = 0 \ \Rightarrow \ (x+9)(x-3) = 0 \ \therefore \ x = -9, \ 3$$

$$\xRightarrow{\text{checking}} \begin{cases} \text{when } x = -9 \ ; \ \sqrt{x+13} - \sqrt{7-x} = \sqrt{4} - \sqrt{16} = 2 - 4 = -2 \text{ (false)} \\ \text{when } x = 3 \ ; \ \sqrt{x+13} - \sqrt{7-x} = \sqrt{16} - \sqrt{4} = 4 - 2 = 2 \text{ (true)} \end{cases}$$

Therefore, $x = 3$ is the solution for the given equation.

11-3 Irrational Numbers

1. Irrational Numbers

(1) Definition

An irrational number is a real number that cannot be expressed as a fraction. It is expressed as the square root of a number that is not a perfect square.

Example $\sqrt{2}$, $\sqrt{3}$, $\sqrt{5}$, $-\sqrt{2}$, $-\sqrt{3}$, $-\sqrt{5}$, $\cdots\cdots$ are irrational numbers.

Note: *Using a calculator, you can find the square roots of numbers.*

$$\sqrt{1} = 1 \quad \sqrt{2} = 1.4142\cdots \quad \sqrt{3} = 1.7320\cdots \quad \sqrt{4} = 2 \quad \sqrt{5} = 2.2360\cdots \quad \sqrt{6} = 2.4494\cdots$$

$$\sqrt{7} = 2.6457\cdots \quad \sqrt{8} = 2.8284\cdots \quad \sqrt{9} = 3 \quad \sqrt{10} = 3.1622\cdots \quad \sqrt{11} = 3.3166\cdots$$

Not: $\text{Decimal} \begin{cases} \text{Finite} \Rightarrow \text{Terminating Decimal} \Rightarrow \text{Rational Number} \\ \text{Infinite} \begin{cases} \text{Repeating Decimal} \Rightarrow \text{Rational Number} \\ \text{Non} - \text{Repeating Decimal} \Rightarrow \text{Irrational Number} \end{cases} \end{cases}$

2. Operations of Irrational Numbers

(1) Multiplying and Dividing Square Roots

For any $a > 0$, $b > 0$,

$$\sqrt{a} \cdot \sqrt{b} = \sqrt{ab} \qquad m\sqrt{a} \cdot n = mn\sqrt{a} \qquad m\sqrt{a} \cdot n\sqrt{b} = mn\sqrt{ab}$$

$$\frac{\sqrt{a}}{\sqrt{b}} = \sqrt{\frac{a}{b}} \qquad m \div \sqrt{a} = \frac{m}{\sqrt{a}} \qquad m\sqrt{a} \div n\sqrt{b} = \frac{m}{n}\sqrt{\frac{a}{b}}$$

1) If the indexes are the same

⇒ Apply the properties of roots.

① $\sqrt[n]{a} \cdot \sqrt[n]{b} = \sqrt[n]{ab}$

② $\dfrac{\sqrt[n]{a}}{\sqrt[n]{b}} = \sqrt[n]{\dfrac{a}{b}}$, $b \neq 0$

Example $\sqrt{3} \cdot \sqrt{5} = \sqrt{3 \cdot 5} = \sqrt{15}$ $\qquad 3\sqrt{2} \cdot 5 = 15\sqrt{2}$ $\qquad 3\sqrt{2} \cdot 4\sqrt{3} = 12\sqrt{6}$

$$\frac{\sqrt{3}}{\sqrt{5}} = \sqrt{\frac{3}{5}} \qquad\qquad 5 \div \sqrt{2} = \frac{5}{\sqrt{2}} \qquad\qquad 3\sqrt{2} \div 4\sqrt{3} = \frac{3}{4}\sqrt{\frac{2}{3}}$$

$$\sqrt[3]{2} \cdot \sqrt[3]{5} = \sqrt[3]{10} \qquad\qquad \frac{\sqrt[3]{2}}{\sqrt[3]{5}} = \sqrt[3]{\frac{2}{5}}$$

2) If the indexes are different

$\Rightarrow$ Rewrite each radical as an exponential expression and apply the properties of exponents.

① $\sqrt[m]{a} \cdot \sqrt[n]{a} = a^{\frac{1}{m}} \cdot a^{\frac{1}{n}} = a^{\frac{1}{m}+\frac{1}{n}}$

② $\dfrac{\sqrt[m]{a}}{\sqrt[n]{a}} = \dfrac{a^{\frac{1}{m}}}{a^{\frac{1}{n}}} = a^{\frac{1}{m}-\frac{1}{n}}$

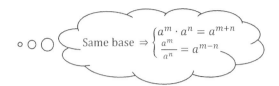

Same base $\Rightarrow \begin{cases} a^m \cdot a^n = a^{m+n} \\ \dfrac{a^m}{a^n} = a^{m-n} \end{cases}$

Example

$\sqrt[3]{5} \cdot \sqrt{5} = 5^{\frac{1}{3}} \cdot 5^{\frac{1}{2}} = 5^{\frac{1}{3}+\frac{1}{2}} = 5^{\frac{5}{6}}$

$\dfrac{\sqrt[3]{5}}{\sqrt{5}} = \dfrac{5^{\frac{1}{3}}}{5^{\frac{1}{2}}} = 5^{\frac{1}{3}-\frac{1}{2}} = 5^{-\frac{1}{6}} = \dfrac{1}{5^{\frac{1}{6}}}$

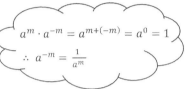

$a^m \cdot a^{-m} = a^{m+(-m)} = a^0 = 1$

$\therefore a^{-m} = \dfrac{1}{a^m}$

$\sqrt{4} \cdot \sqrt[3]{8} = \sqrt[2]{2^2} \cdot \sqrt[3]{2^3} = (2^2)^{\frac{1}{2}} \cdot (2^3)^{\frac{1}{3}} = 2^1 \cdot 2^1 = 2^{1+1} = 2^2 = 4$

$\dfrac{\sqrt{4}}{\sqrt[3]{8}} = \dfrac{\sqrt[2]{2^2}}{\sqrt[3]{2^3}} = \dfrac{(2^2)^{\frac{1}{2}}}{(2^3)^{\frac{1}{3}}} = \dfrac{2^1}{2^1} = 2^{1-1} = 2^0 = 1$

For any a, b, c,

$a \cdot (b + c) = ab + ac$

$(a + b) \cdot c = ac + bc$

(2) Using the Distributive Property for Square Roots

For any $a > 0$, $b > 0$, $c > 0$,

① $\sqrt{a} \cdot (\sqrt{b} + \sqrt{c}) = \sqrt{ab} + \sqrt{ac}$; $(\sqrt{a} + \sqrt{b}) \cdot \sqrt{c} = \sqrt{ac} + \sqrt{bc}$

② $\sqrt{a} \cdot (\sqrt{b} - \sqrt{c}) = \sqrt{ab} - \sqrt{ac}$; $(\sqrt{a} - \sqrt{b}) \cdot \sqrt{c} = \sqrt{ac} - \sqrt{bc}$

Example

$\sqrt{2} \cdot (\sqrt{3} + \sqrt{5}) = \sqrt{2} \cdot \sqrt{3} + \sqrt{2} \cdot \sqrt{5} = \sqrt{6} + \sqrt{10}$

$(\sqrt{6} - \sqrt{2}) \cdot \sqrt{3} = \sqrt{6} \cdot \sqrt{3} - \sqrt{2} \cdot \sqrt{3} = \sqrt{18} - \sqrt{6} = 3\sqrt{2} - \sqrt{6}$

(3) Simplifying Square Roots

1) Moving inside of the radical

For any $a > 0$, $b > 0$,

$$
\begin{array}{l}
① \quad \sqrt{a^2 b} = \sqrt{a^2} \cdot \sqrt{b} = a\sqrt{b} \\[2mm]
② \quad \sqrt{\dfrac{a}{b^2}} = \dfrac{\sqrt{a}}{\sqrt{b^2}} = \dfrac{\sqrt{a}}{b}
\end{array}
$$

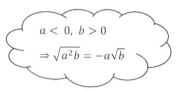

$$a < 0,\ b > 0$$
$$\Rightarrow \sqrt{a^2 b} = -a\sqrt{b}$$

Example

$$\sqrt{12} = \sqrt{2^2 \cdot 3} = \sqrt{2^2} \cdot \sqrt{3} = 2\sqrt{3} \qquad \sqrt{\frac{2}{9}} = \sqrt{\frac{2}{3^2}} = \frac{\sqrt{2}}{\sqrt{3^2}} = \frac{\sqrt{2}}{3}$$

2) Moving outside of the radical

For any $a > 0$, $b > 0$,

$$
\begin{array}{l}
① \quad a\sqrt{b} = \sqrt{a^2} \cdot \sqrt{b} = \sqrt{a^2 b} \\[2mm]
② \quad \dfrac{\sqrt{a}}{b} = \dfrac{\sqrt{a}}{\sqrt{b^2}} = \sqrt{\dfrac{a}{b^2}}
\end{array}
$$

Example

$$3\sqrt{5} = \sqrt{9} \cdot \sqrt{5} = \sqrt{45} \qquad \frac{\sqrt{5}}{3} = \frac{\sqrt{5}}{\sqrt{9}} = \sqrt{\frac{5}{9}}$$

(4) Rationalizing Denominators

To express $\dfrac{1}{\sqrt{2}}$ as a decimal, we have to divide 1 by $\sqrt{2}$.

Since $\dfrac{1}{\sqrt{2}} = \dfrac{1}{1.414\cdots}$, it is quite awkward to calculate. So, you use a simpler procedure to eliminate the radical in the denominator.

That is, $\dfrac{1}{\sqrt{2}} = \dfrac{1 \cdot \sqrt{2}}{\sqrt{2} \cdot \sqrt{2}} = \dfrac{\sqrt{2}}{(\sqrt{2})^2} = \dfrac{\sqrt{2}}{2} \approx \dfrac{1.414}{2} = 0.707$

For any $a > 0$, $b > 0$, $c > 0$,

$$
\begin{array}{l}
① \quad \dfrac{1}{\sqrt{a}} = \dfrac{1 \cdot \sqrt{a}}{\sqrt{a} \cdot \sqrt{a}} = \dfrac{\sqrt{a}}{(\sqrt{a})^2} = \dfrac{\sqrt{a}}{a} \\[4mm]
② \quad \dfrac{\sqrt{b} + \sqrt{c}}{\sqrt{a}} = \dfrac{(\sqrt{b} + \sqrt{c}) \cdot \sqrt{a}}{\sqrt{a} \cdot \sqrt{a}} = \dfrac{\sqrt{ab} + \sqrt{ac}}{a} \\[4mm]
③ \quad \dfrac{1}{\sqrt{a} + \sqrt{b}} = \dfrac{1 \cdot (\sqrt{a} - \sqrt{b})}{(\sqrt{a} + \sqrt{b}) \cdot (\sqrt{a} - \sqrt{b})} = \dfrac{\sqrt{a} - \sqrt{b}}{a - b},\ a \neq b
\end{array}
$$

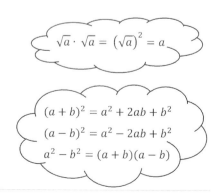

$$\sqrt{a} \cdot \sqrt{a} = \left(\sqrt{a}\right)^2 = a$$

$$(a + b)^2 = a^2 + 2ab + b^2$$
$$(a - b)^2 = a^2 - 2ab + b^2$$
$$a^2 - b^2 = (a + b)(a - b)$$

(5) Adding and Subtracting Square Roots

To add and subtract square roots, the radicands of the square roots must be the same.

1) If the radicands of the square roots are the same, then just add or subtract the numbers which are multiplied by the square roots of the same radicands.

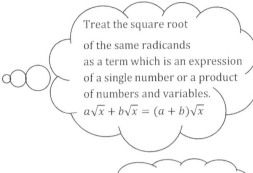

For any real numbers $a(> 0)$, m, and n,

$$m\sqrt{a} + n\sqrt{a} = (m + n)\sqrt{a}$$

$$m\sqrt{a} - n\sqrt{a} = (m - n)\sqrt{a}$$

Treat the square root of the same radicands as a term which is an expression of a single number or a product of numbers and variables.
$a\sqrt{x} + b\sqrt{x} = (a + b)\sqrt{x}$

Example

$$5\sqrt{3} + 4\sqrt{3} = (5 + 4)\sqrt{3} = 9\sqrt{3}$$

$$5\sqrt{3} - 4\sqrt{3} = (5 - 4)\sqrt{3} = \sqrt{3}$$

$\sqrt{a} + \sqrt{b} \neq \sqrt{a + b}$
$\sqrt{a} - \sqrt{b} \neq \sqrt{a - b}$

2) If the radicands of the square roots are different, simplify each of the radicands. If each of the radicands has a perfect square factor, they can be simplified.

Example

$$2\sqrt{12} + 3\sqrt{27} = 2\sqrt{4 \cdot 3} + 3\sqrt{9 \cdot 3} = 2\left(\sqrt{4} \cdot \sqrt{3}\right) + 3\left(\sqrt{9} \cdot \sqrt{3}\right)$$

$$= 2\left(2\sqrt{3}\right) + 3\left(3\sqrt{3}\right) = 4\sqrt{3} + 9\sqrt{3} = (4 + 9)\sqrt{3} = 13\sqrt{3}$$

$\sqrt{a \cdot b} = \sqrt{a} \cdot \sqrt{b}$
$\sqrt{12} = \sqrt{4 \cdot 3} = \sqrt{4} \cdot \sqrt{3} = 2 \cdot \sqrt{3} = 2\sqrt{3}$
$a\sqrt{b}$ means $a \times \sqrt{b}$

(6) Separating Decimal Parts

An irrational number is separated into an integer part and a non-repeating infinite decimal part. Thus, the decimal part can be obtained by removing the integer part from the irrational number.

For any irrational number $A = n + \alpha$ (n is an integer, $0 \leq \alpha < 1$), n is called an integer part and α is called a decimal part of A. The decimal part α of the irrational number A is expressed as $\alpha = A - n$.

For example, $\sqrt{2} = 1.414 \cdots = 1 + 0.414 \cdots = 1 + (\sqrt{2} - 1)$

Example   Separate a number $5 - \sqrt{15}$ into an integer part and a decimal part.

Since $\sqrt{9} = 3$ and $\sqrt{16} = 4$, $3 < \sqrt{15} < 4$ $\therefore \sqrt{15} = 3.\times\times\times$

Since $-4 < -\sqrt{15} < -3$, $5 - 4 < 5 - \sqrt{15} < 5 - 3$ $\therefore 1 < 5 - \sqrt{15} < 2$

Thus, $5 - \sqrt{15} = 1.\times\times\times = 1 + 0.\times\times\times = 1 + $ decimal part

Therefore, the integer part of $5 - \sqrt{15}$ is 1 and

the decimal part is $(5 - \sqrt{15}) - 1 = 4 - \sqrt{15}$.

> For $a > 0$; $\sqrt{a} > 0$:
> $\sqrt{a} = ($ positive integer part$) + ($ decimal part$)$
> $\therefore$ decimal part $= \sqrt{a} - $ positive integer part

Example   Separate a number $5 - 2\sqrt{15}$ into an integer part and a decimal part.

Since $3 < \sqrt{15} < 4$, $6 < 2\sqrt{15} < 8$ $\therefore 2\sqrt{15} = 6.\times\times\times$ or $7.\times\times\times$

In this case, you have to consider another way to solve the problem.

Note that $2\sqrt{15} = \sqrt{4 \cdot 15} = \sqrt{60}$.

Since $7^2 = 49$ and $8^2 = 64$, $7 < 2\sqrt{15} < 8$ $\therefore 2\sqrt{15} = 7.\times\times\times$

Thus, $7 < 2\sqrt{15} < 8$ $\Rightarrow -8 < -2\sqrt{15} < -7$ $\Rightarrow 5 - 8 < 5 - 2\sqrt{15} < 5 - 7$

$\Rightarrow -3 < 5 - 2\sqrt{15} < -2$

$\therefore 5 - 2\sqrt{15} = -2.\times\times\times = -2 - 0.\times\times\times = -2 - ($ decimal part$)$

Therefore, the integer part of $5 - 2\sqrt{15}$ is -2 and

the decimal part is $-2 - (5 - 2\sqrt{15}) = -7 + 2\sqrt{15}$.

> For a negative irrational number :
> negative irrational number $= ($ negative integer part$) - ($ decimal part $)$
> $\therefore$ decimal part $= ($ negative integer part$) - ($ negative irrational number$)$

11-4 Real Numbers

1. Classification of Real Numbers

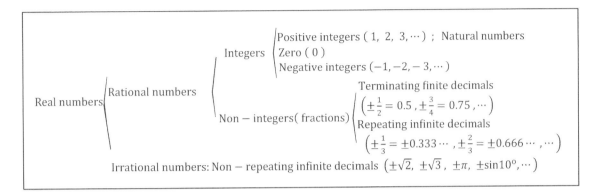

Real numbers
$\Bigg\{$
Rational numbers $\Bigg\{$
Integers $\begin{cases} \text{Positive integers (1, 2, 3,}\cdots\text{)} \; ; \; \text{Natural numbers} \\ \text{Zero (0)} \\ \text{Negative integers } (-1, -2, -3, \cdots) \end{cases}$

Non $-$ integers(fractions) $\begin{cases} \text{Terminating finite decimals} \\ \left(\pm\frac{1}{2} = 0.5 \,, \pm\frac{3}{4} = 0.75 \,, \cdots \right) \\ \text{Repeating infinite decimals} \\ \left(\pm\frac{1}{3} = \pm0.333\cdots \,, \pm\frac{2}{3} = \pm0.666\cdots \,, \cdots \right) \end{cases}$

Irrational numbers: Non $-$ repeating infinite decimals $\left(\pm\sqrt{2}, \; \pm\sqrt{3}, \; \pm\pi, \; \pm\sin10^{\circ}, \cdots \right)$

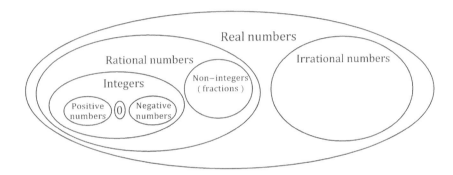

2. Geometric Representation of Real Numbers

Note: *The Pythagorean Theorem*

   The square of the length of a hypotenuse of a right triangle is equal to the sum of the squares of the

   lengths of the legs. That is, if a and b are the lengths of the two legs in a right triangle, and c is the

   length of the hypotenuse, then $a^2 + b^2 = c^2$

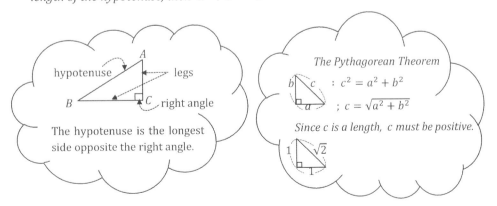

hypotenuse   legs

right angle

The hypotenuse is the longest
side opposite the right angle.

The Pythagorean Theorem

$: c^2 = a^2 + b^2$

$; c = \sqrt{a^2 + b^2}$

Since c is a length, c must be positive.

(1) Representation of Real Numbers as Points on a Line

1) Using the relationship between the length and the area of a square, irrational numbers can appear on a number line. Therefore, there are infinite real numbers between two different real numbers on a number line.

2) Every point on a line corresponds to exactly one real number, and that real number corresponds to one, and only one, point.

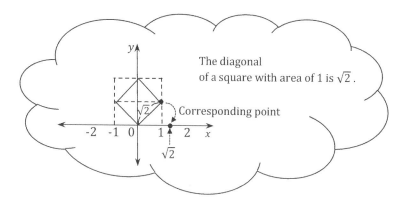

The diagonal of a square with area of 1 is $\sqrt{2}$.

Corresponding point

(2) Magnitude of Real Numbers

For any real numbers a and b, $\quad a > b$ or $a < b$ or $a = b$.

1) $a - b > 0$ $(a - b$ is positive$)$ $\Rightarrow$ $a > b$ $(a$ is greater than $b)$

2) $a - b < 0$ $(a - b$ is negative$)$ $\Rightarrow$ $a < b$ $(a$ is less than $b)$

3) $a - b \geq 0$ $\Rightarrow$ $a \geq b$ $(a$ is greater than or equal to $b)$

4) $a - b \leq 0$ $\Rightarrow$ $a \leq b$ $(a$ is less than or equal to $b)$

For any rational numbers a, b, c, and d,

$m > 0$ $\Rightarrow$

      1) $a + b\sqrt{m} = 0$ $\Leftrightarrow$ $a = 0$ and $b = 0$

      2) $a + b\sqrt{m} = c + d\sqrt{m}$ $\Leftrightarrow$ $a = c$ and $b = d$

$m > 0$, $n > 0$ $\Rightarrow$

      3) $a + \sqrt{m} = b + \sqrt{n}$ $\Leftrightarrow$ $a = b$ and $m = n$

(3) Correspondence between Real Numbers and Points on a Line

1) Coordinate

The coordinate is the number associated with a point. We use coordinates to identify points.

The length between two points is always a positive number.

Thus, we use the absolute value to express the length.

2) Length (Distance)

The *length* (whose end points are a and b) is defined by the real number $|b - a|$.

By the Pythagorean Theorem,

the *length* (*distance*) between two points, $a(x_1, y_1)$ and $b(x_2, y_2)$, is

$$d = \sqrt{(x_2 - x_1)^2 + (y_2 - y_1)^2}$$

3) Midpoint

The coordinates of the midpoint of a line segment are the average values of the corresponding coordinates of the two endpoints.

The *midpoint* m of the line segment joining the points $a(x_1, y_1)$ and $b(x_2, y_2)$ in the coordinate plane is $m = \left(\frac{x_1 + x_2}{2}, \frac{y_1 + y_2}{2} \right)$.

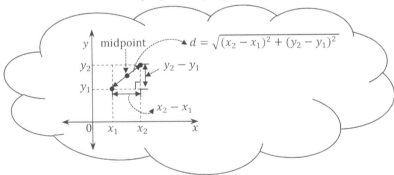

3. Identity and Inverse

For any real number a,

(1) $a + 0 = a$: The value of any real number is not changed by adding 0 to that number.

0 is called the *additive identity*.

(2) $a \times 1 = a$: The value of any real number is not changed by multiplying 1 to that number.

1 is called the *multiplicative identity*.

(3) $a + (-a) = 0$: The sum of any real number and its opposite is always 0.

$-a$ is called the *additive inverse* of a.

Note : If a is a positive integer (natural number), the additive inverse of a is not −a.

This is because −a (negative integer) is not included in positive integers.

However, for any integer a; rational number a ; real number a, −a is the additive inverse of a .

---

(4) $a \times \dfrac{1}{a} = 1, \; a \neq 0$: The product of any real number (except 0) and its reciprocal is always 1.

The real number $\dfrac{1}{a} \; (a \neq 0)$ is called the *multiplicative inverse* of the

real number $a (\neq 0)$.

Note : ① $a \times b = 1 \Rightarrow$ *a and b are reciprocals of each other.*

② The reciprocal of a fraction is obtained by reversing the numerator and denominator.

For any $a(\neq 0)$ and $b(\neq 0)$, the reciprocal of $\dfrac{a}{b}$ is $\dfrac{b}{a}$.

For example, the reciprocal of a is $\dfrac{1}{a}$, because $a = \dfrac{a}{1}$.

③ *If $a(\neq 0)$ is an integer, $\dfrac{1}{a}$ is not the reciprocal of a .*

This is because $\dfrac{1}{a}$ (fraction) is not included in integers.

Thus, there is no reciprocal of a in integers.

However, $\dfrac{1}{a}$ is the reciprocal of a in rational numbers (fractions) and real numbers.

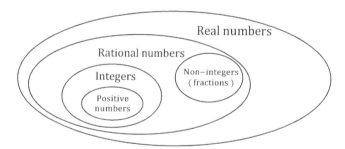

4. Absolute Values

(1) Minus Sign

1) $-a$: the additive inverse of a

2) $a - b$: the difference between a and b

(2) Definition

The *absolute value* of a real number a, denoted by $|a|$, is the real number such that:

$$|a| = \begin{cases} a , & \text{if } a \geq 0 \\ -a , & \text{if } a < 0 \end{cases}$$

The absolute values of all the real numbers except 0 are positive.

Example $|5| = 5, \; |0| = 0, \; |-5| = -(-5) = 5$

(3) Properties of Absolute Value

For any real numbers $a(\neq 0)$ and $b(\neq 0)$,

1) If a and b have the same sign,

$$a + b = \begin{cases} |a| + |b|, & \text{if } a > 0 \text{ and } b > 0 \\ -(|a| + |b|), & \text{if } a < 0 \text{ and } b < 0 \end{cases}$$

2) If a and b have different signs and $|a| > |b|$,

$$a + b = \begin{cases} |a| - |b|, & \text{if } a > 0 \\ -(|a| - |b|), & \text{if } a < 0 \end{cases}$$

Example

$2 + 3 = |2| + |3| = 5$

$(-2) + (-3) = -(|-2| + |-3|) = -(2 + 3) = -5$

$5 + (-4) = |5| - |-4| = 5 - 4 = 1$

$(-5) + 4 = -(|-5| - |4|) = -(5 - 4) = -1$

Note: On a number line, the distance between two points is always a positive number.

So, we use the absolute value for the distance.

And absolute value is a geometric representation of real numbers.

$$\begin{array}{ccccccc} -2 & -1 & 0 & 1 & 2 & 3 & \textit{Number line} \end{array}$$

The distance between -1 and 2 is $|-1 - 2| = |-3| = 3$ or $|2 - (-1)| = |3| = 3$

5. Formal Properties of Real Numbers

For any real numbers a, b, and c,

Reflexive property :	$a = a$
Symmetric property :	$a = b \Rightarrow b = a$
Transitive property :	$a = b$ and $b = c \Rightarrow a = c$

(1) Addition

① Closure property

$a + b$ is a unique real number.

② Commutative property

$a + b = b + a$

③ Associative property

$$(a + b) + c = a + (b + c)$$

④ Identity property

$$a + 0 = 0 + a = a$$

⑤ Inverse property

$$a + (-a) = (-a) + a = 0$$

(2) Multiplication

① Closure property

$$a \cdot b \text{ is a unique real number.}$$

② Commutative property

$$a \cdot b = b \cdot a$$

③ Associative property

$$(a \cdot b) \cdot c = a \cdot (b \cdot c)$$

④ Identity property

$$a \cdot 1 = 1 \cdot a = a$$

⑤ Inverse property

$$a \cdot \frac{1}{a} = \frac{1}{a} \cdot a = 1, \ a \neq 0$$

⑥ Distributive property

$$a \cdot (b + c) = (a \cdot b) + (a \cdot c), \qquad (a + b) \cdot c = (a \cdot c) + (b \cdot c)$$

(3) Equality

$$a = b \ \Rightarrow \ \text{①} \ a + c = b + c \quad \text{(Addition)}$$
$$\text{②} \ a - c = b - c \quad \text{(Subtraction)}$$
$$\text{③} \ a \cdot c = b \cdot c \quad \text{(Multiplication)}$$
$$\text{④} \ \frac{a}{c} = \frac{b}{c}, \ c \neq 0 \quad \text{(Division)}$$

6. Exponents

For any real numbers $a(\neq 0)$, $b(\neq 0)$, and positive integers m and n, the following rules apply:

(1) Addition of exponent

$$a^m \cdot a^n = a^{m+n} \ ; \quad a^{m+1} - a^m = a^m \cdot a^1 - a^m = a^m(a - 1)$$

(2) Multiplication of exponent

$$(a^m)^n = a^{mn}$$

(3) Division of exponent

$$\frac{a^m}{a^n} = a^{m-n} \quad \text{or} \quad \frac{a^m}{a^n} = \frac{1}{a^{n-m}}$$

(4) Negative exponent

$$\frac{1}{a^m} = a^{-m}, \quad \frac{1}{a^{-m}} = a^m \quad ; \quad a^m \cdot a^{-m} = 1 \quad ; \quad a^0 = 1$$

(5) Distributive property of exponent

$$(ab)^m = a^m \cdot b^m$$

$$\left(\frac{a}{b}\right)^m = \frac{a^m}{b^m}$$

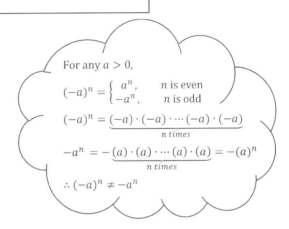

For any $a > 0$,

$$(-a)^n = \begin{cases} a^n, & n \text{ is even} \\ -a^n, & n \text{ is odd} \end{cases}$$

$$(-a)^n = \underbrace{(-a) \cdot (-a) \cdots (-a) \cdot (-a)}_{n \ times}$$

$$-a^n = -\underbrace{(a) \cdot (a) \cdots (a) \cdot (a)}_{n \ times} = -(a)^n$$

$$\therefore (-a)^n \neq -a^n$$

Exercises

#1   Simplify the following exponents

(1) $2^5 \cdot 2^7$

(2) $3^{12} \cdot 3^{15}$

(3) $(3^4)^5$

(4) $(2^3)^5 \cdot 2^3$

(5) $(4^5)^{6+1}$

(6) $2^3 \cdot 4^5$

(7) $4^3 \cdot 16^2$

(8) $(2 \cdot 8)^3$

(9) $(3 \cdot 4)^5$

(10) $4^2 \cdot 12^3$

(11) $3^{-2} \cdot 3^5 \cdot 3^0$

(12) $2^3 \cdot 5^{-2} \cdot 2^{-6} \cdot 5^3$

(13) $(-2)^3 \cdot 4^3 \cdot 2^{-3}$

(14) $\frac{2 \cdot 5^{-1} + 1 + 2^{-1} \cdot 5}{2^{-1} \cdot 5^{-1}}$

(15) $(-2)^4 \cdot 4^{-2} \cdot (-2)^3 \cdot 8^{-1}$

#2   Find the square roots of the following

(1) 36

(2) 9

(3) 25

(4) 49

(5) 121

(6) $\frac{9}{16}$

(7) 0.04

(8) 225

(9) 0.16

(10) $\frac{36}{49}$

#3   Find the value of each square root.

(1) $\sqrt{36}$

(2) $\sqrt{81}$

(3) $\sqrt{100}$

(4) $\sqrt{1}$

(5) $\sqrt{0.49}$

(6) $\sqrt{\frac{4}{16}}$

(7) $\sqrt{0}$

(8) $\sqrt{169}$

(9) $\sqrt{0.01}$

(10) $\sqrt{\frac{9}{144}}$

#4   Compare the square of the numbers using the signs $>$ or $<$.

(1) 2 and $\sqrt{3}$

(2) $\sqrt{2}$ and 1.3

(3) $\sqrt{3}$ and $\frac{3}{4}$

(4) $\sqrt{5}$ and 2.5

(5) $-\sqrt{\dfrac{4}{3}}$ and $-\sqrt{\dfrac{3}{2}}$

(6) $\sqrt{0.5}$ and $\sqrt{0.05}$

(7) $-\sqrt{2}$ and $\sqrt{5}$

(8) $-\sqrt{0.3}$ and $-\sqrt{\dfrac{3}{4}}$

(9) $\sqrt{2}$, 1.4, and 1.5

(10) $-\sqrt{3}$, -1.6, and -1.8

#5 Simplify the following square roots

(1) $\sqrt{4 \cdot 9}$

(2) $\sqrt{24}$

(3) $\sqrt{3}\,(\sqrt{27})$

(4) $\sqrt{2}\,(\sqrt{30})$

(5) $\sqrt{2^3 \cdot 3^4 \cdot 5^3}$

(6) $\sqrt{280}$

(7) $\sqrt{216}$

(8) $\sqrt{10} \cdot \sqrt{12} \cdot \sqrt{6}$

(9) $\sqrt{\dfrac{6}{10}} \cdot \sqrt{\dfrac{5}{14}}$

(10) $\sqrt{\dfrac{5}{8}} \cdot \sqrt{\dfrac{3}{10}} \cdot \sqrt{\dfrac{4}{9}}$

(11) $\dfrac{\sqrt{15}}{\sqrt{3}}$

(12) $\sqrt{18} \div \sqrt{8}$

(13) $\dfrac{5\sqrt{45}}{3\sqrt{3}}$

(14) $\dfrac{\sqrt{42}}{\sqrt{56}} \cdot \sqrt{24}$

(15) $5\sqrt{\dfrac{3}{4}} \cdot 2\sqrt{\dfrac{24}{36}}$

#6 Solve the following radical equations

(1) $\sqrt{x+5} = 3$

(2) $\sqrt{x^2 + 2x} - 2x = 0$

(3) $\sqrt{x+1} + \sqrt{x+5} = 0$

(4) $\sqrt{x+11} - \sqrt{x+18} = -1$

#7 Simplify the following square roots

(1) $\sqrt{3} + 5\sqrt{3}$

(2) $\sqrt{12} + \sqrt{27}$

(3) $\sqrt{45} + \sqrt{20}$

(4) $\sqrt{24} - \sqrt{54}$

(5) $\sqrt{18} - \sqrt{50}$

(6) $\sqrt{3} + 3\sqrt{27} - 2\sqrt{48}$

(7) $\sqrt{2}(\sqrt{10} + \sqrt{18})$

(8) $(\sqrt{8} + 3\sqrt{3})\,\sqrt{2}$

(9) $\sqrt{5}(\sqrt{12} - 3) + \sqrt{32}$

(10) $(\sqrt{2} + 2)(\sqrt{6} + \sqrt{3})$

(11) $(\sqrt{5} - 3)(\sqrt{6} - \sqrt{3})$

(12) $(\sqrt{3} + 3)(\sqrt{3} - 3)$

(13) $(2\sqrt{5} - \sqrt{2})(2\sqrt{5} + \sqrt{2})$

(14) $(\sqrt{5} + \sqrt{2})^2$

(15) $2\sqrt{5}(3\sqrt{3} - \sqrt{2})^2$

(16) $(\sqrt{3} - 2\sqrt{2})^2 + (\sqrt{3} + 2\sqrt{2})^2$

(17) $(1 + \sqrt{2} - \sqrt{3})(1 - \sqrt{2} - \sqrt{3})$

(18) $(\sqrt{2} + \sqrt{3})(\sqrt{2} + \sqrt{3} - \sqrt{5})$

(19) $\sqrt{2} \cdot \sqrt[3]{16}$

(20) $\sqrt[3]{20} \cdot \sqrt{15}$

(21) $\dfrac{\sqrt[3]{8}}{\sqrt{8}}$

(22) $\dfrac{\sqrt{15}}{\sqrt{5}} \div \dfrac{5}{3\sqrt{2}}$

(23) $\sqrt{3} \div \sqrt{18} \div \sqrt{3} \cdot \sqrt{32}$

(24) $\sqrt{\dfrac{3}{4}} \div \sqrt{\dfrac{15}{10}} \div \dfrac{1}{\sqrt{6}}$

#8 Rationalize the denominators for the following

(1) $\dfrac{3}{\sqrt{2}}$

(2) $\dfrac{5}{\sqrt{3}}$

(3) $\dfrac{\sqrt{18}}{\sqrt{12}}$

(4) $\sqrt{\dfrac{4}{3}}$

(5) $5\sqrt{\dfrac{1}{45}}$

(6) $\dfrac{3}{2\sqrt{3}} - 4\sqrt{3}$

(7) $\sqrt{\dfrac{2}{3}} - \sqrt{\dfrac{5}{18}}$

(8) $\dfrac{3\sqrt{5} - \sqrt{3}}{\sqrt{3}}$

(9) $\sqrt{\dfrac{2}{3}} - \sqrt{5} + \sqrt{\dfrac{3}{2}}$

(10) $\dfrac{2\sqrt{3}}{\sqrt{3} - \sqrt{2}} + \dfrac{3\sqrt{2}}{\sqrt{3} + \sqrt{2}}$

#9 Find the value of $a + b$, $a - b$, ab and $a^2 + b^2$ for the following given expressions

(1) $a = \sqrt{3} + \sqrt{5}$, $b = \sqrt{3} - \sqrt{5}$

(2) $a = \dfrac{1}{\sqrt{3} + \sqrt{5}}$, $b = \dfrac{1}{\sqrt{3} - \sqrt{5}}$

(3) $a = \dfrac{\sqrt{3} - \sqrt{5}}{\sqrt{3} + \sqrt{5}}$, $b = \dfrac{\sqrt{3} + \sqrt{5}}{\sqrt{3} - \sqrt{5}}$

#10 Find the value of each square root.

(1) $\sqrt{(-5)^2}$

(2) $-\sqrt{5^2}$

(3) $\sqrt{(1 - 4)^2}$

(4) $\sqrt[3]{(-5)^2}$

(5) $\sqrt[3]{(-5)^3}$

(6) $\sqrt{(-2)^2} + \sqrt[3]{(-2)^3} - 5\sqrt{(-2)^2}$

(7) $\sqrt[3]{(-5)^3} - \sqrt{(-5)^2} + \sqrt{5^3}$

(8) $\dfrac{\sqrt[3]{20}}{\sqrt[3]{10}}$

(9) $\dfrac{\sqrt[3]{20}}{\sqrt{10}}$

(10) $3\sqrt[4]{4^2} - \sqrt[4]{(-2)^4}$

#11 For rational numbers a and b, each expression is a rational number. Find the value of ab.

(1) $\dfrac{3+b\sqrt{2}}{\sqrt{2}-a}$

(2) $\dfrac{\sqrt{3}+b}{a\sqrt{3}+2}$

#12 Solve the following

(1) $a = \dfrac{3+\sqrt{2}}{3-\sqrt{2}}$ Find the value of $a + \dfrac{1}{a}$ and $a - \dfrac{1}{a}$.

(2) $a - \dfrac{1}{a} = 2\sqrt{3} - 5$ Find the value of $\left(a + \dfrac{1}{a}\right)^2$.

(3) $a = 2 - 3\sqrt{2} + \sqrt{5}$ Find the value of $a^2 - 4a$.

#13 Simplify the following expressions

(1) $\sqrt{2^2} - \left(-\sqrt{2^2}\right)^2 \cdot 2 + \sqrt{(-3)^2} \cdot (-3^2)$

(2) $\sqrt{(-3)^2} \cdot \sqrt{(-1)^2} \div \left(-\sqrt{3}\right)^2 + \sqrt{2} \cdot \left(-\sqrt{2}\right)^2$

(3) $a > 0$, $\sqrt{4a^2} - \sqrt{(-a)^2}$

(4) $a < 0$, $\sqrt{(-a)^2} - \sqrt{(-3a)^2} - \sqrt{(-2a)^2}$

(5) $0 < a < 1$, $\sqrt{a^2} - \sqrt{(a-1)^2}$

(6) $1 < a < 3$, $\sqrt{(-a)^2} - \sqrt{(a-1)^2} + \sqrt{16(a-3)^2}$

(7) $\sqrt{(\sqrt{2}-1)^2} - \sqrt{\left(1-\sqrt{2}\right)^2} - \sqrt{\left(-\sqrt{2}\right)^2}$

(8) $\sqrt{(\sqrt{11}-3)^2} - \sqrt{\left(2-\sqrt{11}\right)^2}$

(9) $a < b < c$, $\sqrt{(a-b)^2} - \sqrt{(b-c)^2} - \sqrt{(c-a)^2}$

(10) $a + b < 0$, $ab > 0$, $\sqrt{(-a)^2} - \sqrt{(-b)^2} - \sqrt{(-2a)^2} + \sqrt{(-2b)^2}$

#14 Solve the following

(1) a is the integer part of $3\sqrt{5}$ and b is the decimal part of $2\sqrt{7}$.

Find the value of $a + b$ and $a - b$.

(2) a and b are the integer part and the decimal part of $7 - \sqrt{12}$, respectively.

Find the value of ab and $\dfrac{a}{b}$.

#15 In each of following, determine whether the statement is true or false.

(1) Infinite decimals are irrational numbers.

(2) All square roots of a number are irrational numbers.

(3) Repeating decimals are rational numbers.

(4) Non-repeating infinite decimals are irrational numbers.

(5) A number using a radical sign is an irrational number.

(6) There are numbers that are both rational and irrational.

(7) Irrational numbers cannot be expressed as fractions.

(8) 0 is neither a rational number nor an irrational number.

(9) Irrational numbers are positive numbers.

(10) There are no irrational numbers between $\sqrt{2}$ and $\sqrt{3}$.

#16 Find the distance (ab) and midpoint (m) between each pair of points with the following coordinates

(1) $a(1, 2)$, $b(3, 4)$

(2) $a(-2, 3)$, $b(3, -5)$

(3) $a(3, 4)$, $b(-2, -3)$

(4) $a(-\sqrt{2}, 3)$, $b(3, \sqrt{2})$

(5) $a(0, -1)$, $b\left(-4, \frac{1}{2}\right)$

#17 $a = -2$, $b = -\frac{1}{3}$, $c = 4$ Simplify the following expressions

(1) $|a - b|$

(2) $|a| - |b|$

(3) $|c + (-b)|$

(4) $|a| + |b| - |c|$

(5) $|a - c| - |b|$

(6) $a - |b - c|$

(7) $-c - |a + b|$

(8) $|-b + c| - |a|$

(9) $|a| - |-b| - |-c|$

(10) $|a - b - c|$

#18 Find all the integers satisfying the following inequalities

(1) $1 < \sqrt{|x - 2|} < \sqrt{5}$

(2) $\sqrt{6} < \sqrt{\left|\frac{x-1}{2}\right|} < 3$

#19   Simplify the following expression for $a < 0$ and $b < 0$

$$\sqrt{(-a)^2} + |b| - \sqrt{(-b)^2} - 2|a + b|$$

#20   Simplify the following expressions Type equation here.

(1) $\dfrac{|2-4|}{|-3\sqrt{2}|-2}$

(2) $\left| \dfrac{-|\sqrt{2}-\sqrt{5}|}{\sqrt{5}-\sqrt{2}} \right|$

Chapter 12. Factorization

12-1 Factoring Polynomials

1. Definition

(1) The Greatest Common Factor (GCF)

If a polynomial $P(x)$ is expressed as

$P(x) = Q_1(x) \cdot Q_2(x) \cdots Q_n(x)$ where $Q_1(x), Q_2(x), \cdots, Q_n(x)$ are polynomials, then $Q_1(x), Q_2(x), \cdots, Q_n(x)$ are called factors of $P(x)$.

The Greatest Common Factor (GCF) is the product of their common factors raised to their lowest powers.

Example

$90x^2y^3 + 12x^4y^5$ It cannot be factored further.

Since $90 = 2 \cdot 3^2 \cdot 5$ (prime factorization) and $12 = 2^2 \cdot 3$ (prime factorization), the common factors are $2, \ 3, x, \ $ and $ \ y$.

$\therefore \ \text{GCF} = \ 2^1 \cdot 3^1 \cdot x^2 \cdot y^3 = 6x^2y^3$

(2) Factorization

Factorization is an expression of a polynomial as a product of its greatest common factor and two or more prime polynomials.

Example

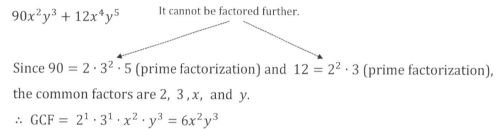

$3a^2 + 15a = \underline{3a}\,((a + 5))$ Remaining factor when the GCF is factored out of the polynomial.

GCF of $3a^2$ and $15a$

Example

$2a(x - 3) - 4(x - 3) = 2\,(x - 3)((a - 2))$

GCF of $2a(x - 3)$ and $4(x - 3)$

Note

$(x + 2)(x + 3) = x^2 + 5x + 6$ *Multiplying two polynomials using the distributive property*

(Expanding)

$x^2 + 5x + 6 = (x + 2)(x + 3)$ *Factoring polynomials as a product of two prime polynomials*

Note: The relationship of GCF and LCM

Let A and B be two polynomials whose leading coefficients are 1.

If GCF of A and B is G and LCM of A and B is L, then

1) $A = aG, B = bG$

2) $L = abG = aB = bA$

3) $AB = abG^2 = LG$

4) $A + B = (a + b)G, \quad A - B = (a - b)G$

for any a and b which don't have any common factor except 1.

12-2 Factorization Formulas

- *Quadratic: The highest power in the polynomial is 2.*
- *Trinomial: A polynomial containing 3 terms*
- *Quadratic trinomial is the form of $ax^2 + bx + c, \ a \neq 0$*

1. The Form of a Perfect Square

$$a^2 + 2ab + b^2 = (a + b)^2$$

$$a^2 - 2ab + b^2 = (a - b)^2$$

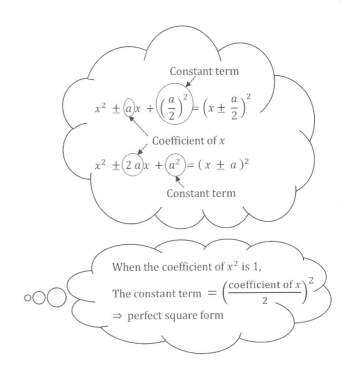

Example

$$x^2 + \fbox{6}x + \fbox{9} = (x + 3)^2$$

$$2 \cdot 3 \qquad 3^2$$

$$x^2 + \fbox{5}x + \fbox{6} \neq (\cdots)^2$$

$$2 \cdot \frac{5}{2} \qquad 6 \neq \left(\frac{5}{2}\right)^2$$

Note

1) $a - b = -(b - a)$

2) $(a - b)^2 = (-(b - a))^2 = (b - a)^2$

3) $(a - b)^3 = (-(b - a))^3 = -(b - a)^3$

4) $(a - b)^4 = (-(b - a))^4 = (b - a)^4$

2. The Form of $(a + b)(a - b)$

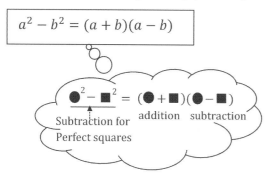

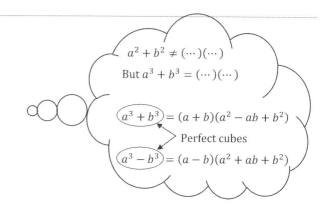

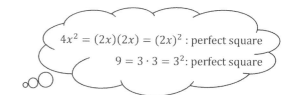

Note:

$$-a^2 + b^2 = b^2 - a^2 = (b - a)(b + a)$$

Example

$$4x^2 - 9 = (2x)^2 - (3)^2 = (2x + 3)(2x - 3)$$

3. Quadratic Trinomials with Binomial Factors

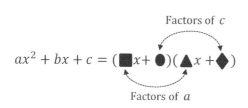

(1) If the Leading Coefficient is 1,

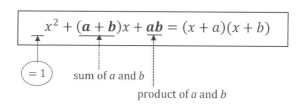

① Step 1   Find all the possible numbers which factor the constant ab.

② Step 2   Check to see if the sum of the two numbers found in Step 1 is the same as the coefficient of x.

③ Step 3   Factor $(x + a)(x + b)$.

Example

$$x^2 + \boxed{3}x + \boxed{2} = (x + 1)(x + 2)$$

$1 + 2 \qquad 1 \times 2$

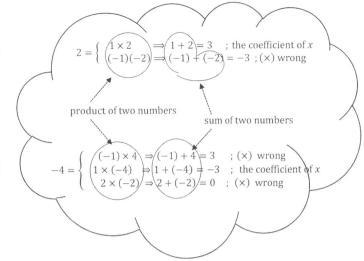

$$2 = \begin{cases} 1 \times 2 & \Rightarrow 1 + 2 = 3 & ; \text{ the coefficient of } x \\ (-1)(-2) & \Rightarrow (-1) + (-2) = -3 & ; (\times) \text{ wrong} \end{cases}$$

product of two numbers

sum of two numbers

Example

$$x^2 \boxed{-3}x \boxed{-4} = (x - 4)(x + 1)$$

$-4 + 1 \qquad -4 \times 1$

$$-4 = \begin{cases} (-1) \times 4 & \Rightarrow (-1) + 4 = 3 & ; (\times) \text{ wrong} \\ 1 \times (-4) & \Rightarrow 1 + (-4) = -3 & ; \text{ the coefficient of } x \\ 2 \times (-2) & \Rightarrow 2 + (-2) = 0 & ; (\times) \text{ wrong} \end{cases}$$

(2) If The Leading Coefficient is not 1,

$$\boxed{\underline{ac}x^2 + (ad + bc)x + bd = (ax + b)(cx + d)}$$

$\neq 1$

Note:

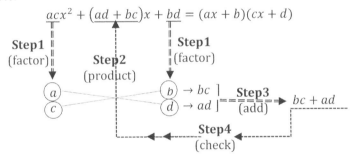

$$acx^2 + (ad + bc)x + bd = (ax + b)(cx + d)$$

Step1
(factor)

Step2
(product)

Step1
(factor)

a

c

$b \rightarrow bc$

$d \rightarrow ad$

Step3
(add)

$bc + ad$

Step4
(check)

① Step 1. Find all the possible numbers which factor the coefficient of x^2 and the constant

② Step 2. Product the numbers diagonally

③ Step 3. Add the numbers obtained from step 2

④ Step 4. Check to see if the sum obtained from Step 3 is the same as the coefficient of x

⑤ Step 5. Factor as $(ax + b)(cx + d)$

Example

$$2x^2 + 7x + 6 = (x + 2)(2x + 3)$$

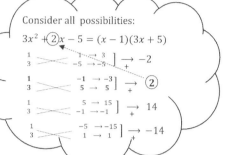

$$
\begin{array}{l}
1 \\
2
\end{array}
\times
\begin{array}{ll}
2 & \to \quad 4 \\
3 & \to \quad 3
\end{array}
\Big] \xrightarrow[+]{} 7
$$

> $2 = \mathbf{1} \times \mathbf{2}$ or 2×1
> or $(-1) \times (-2)$
> or $(-2) \times (-1)$
> $6 = 1 \times 6$ or 6×1
> or $\mathbf{2} \times \mathbf{3}$ or 3×2
> or $(-1) \times (-6)$ or $(-6) \times (-1)$
> or $(-2) \times (-3)$ or $(-3) \times (-2)$

Example

$$3x^2 + 2x - 5 = (x - 1)(3x + 5)$$

$$
\begin{array}{l}
1 \\
3
\end{array}
\times
\begin{array}{ll}
-1 & \to \quad -3 \\
5 & \to \quad 5
\end{array}
\Big] \xrightarrow[+]{} 2
$$

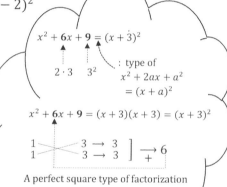

> Consider all possibilities:
> $3x^2 + \textcircled{2}x - 5 = (x - 1)(3x + 5)$
> $\begin{array}{l} 1 \\ 3 \end{array} \times \begin{array}{ll} 1 & \to \ 3 \\ -5 & \to -5 \end{array} \Big] \xrightarrow[+]{} -2$
> $\begin{array}{l} 1 \\ 3 \end{array} \times \begin{array}{ll} -1 & \to \ -3 \\ 5 & \to \ 5 \end{array} \Big] \xrightarrow[+]{} \textcircled{2}$
> $\begin{array}{l} 1 \\ 3 \end{array} \times \begin{array}{ll} 5 & \to \ 15 \\ -1 & \to -1 \end{array} \Big] \xrightarrow[+]{} 14$
> $\begin{array}{l} 1 \\ 3 \end{array} \times \begin{array}{ll} -5 & \to -15 \\ 1 & \to \ 1 \end{array} \Big] \xrightarrow[+]{} -14$

Example

$$9x^2 - 12x + 4 = (3x - 2)(3x - 2) = (3x - 2)^2$$

$$
\begin{array}{l}
3 \\
3
\end{array}
\times
\begin{array}{ll}
-2 & \to \quad -6 \\
-2 & \to \quad -6
\end{array}
\Big] \xrightarrow[+]{} -12
$$

> $x^2 + 6x + 9 = (x + 3)^2$
> $2 \cdot 3 \quad 3^2$: type of
> $x^2 + 2ax + a^2$
> $= (x + a)^2$
> $x^2 + 6x + 9 = (x + 3)(x + 3) = (x + 3)^2$
> $\begin{array}{l} 1 \\ 1 \end{array} \times \begin{array}{ll} 3 & \to \ 3 \\ 3 & \to \ 3 \end{array} \Big] \xrightarrow[+]{} 6$
> A perfect square type of factorization is a special case for factoring quadratic trinomials.

12-3 More Factorization

1. If each term has common factors,

⇒ Find the GCF first,

express the polynomial as a product of the GCF,

and prime polynomials.

Note

$$ax^2 + 4ax + 4a$$
$$\xRightarrow[\text{find the GCF}]{} a(x^2 + 4x + 4)$$
$$\xRightarrow[\text{factor}]{} a(x + 2)^2$$

Example

$$2x^2 + 6x + 4$$

$$
\begin{array}{l}
1 \\
2
\end{array}
\times
\begin{array}{ll}
2 & \to \quad 4 \\
2 & \to \quad 2
\end{array}
\Big] \xrightarrow[+]{} 6
$$

$$= (x + 2)(2x + 2) = 2(x + 2)(x + 1)$$

Begin by removing a common factor, and then factoring further.

That is, $2x^2 + 6x + 4 = 2(x^2 + 3x + 2)$

$$x^2 + 3x + 2$$

$$
\begin{array}{l}
1 \quad \rightarrow \quad 1 \\
1 \quad \rightarrow \quad 1 \\
1 \quad \quad 2 \rightarrow 2
\end{array}
\Bigg] \xrightarrow[+]{} 3
$$

> Common factors should be removed before other methods of factoring are used.

$$= (x + 1)(x + 2)$$

$$\therefore \ 2x^2 + 6x + 4 = 2(x^2 + 3x + 2) = 2(x + 1)(x + 2)$$

2. If a polynomial contains common terms,

$\Rightarrow$ Substitute the common terms with other variables. Then factor the resulting polynomial.

Rewrite it with common terms.

Example

$$
\begin{aligned}
(a - 1)^2 + 2(a - 1) - 3 &= A^2 + 2A - 3 \quad \text{Substitute } a - 1 \text{ as } A \\
&= (A + 3)(A - 1) \quad \text{Factor} \\
&= (a - 1 + 3)(a - 1 - 1) \quad \text{Replace } A \text{ with } a - 1 \\
&= (a + 2)(a - 2)
\end{aligned}
$$

3. If polynomials containing more than 3 terms have no common factors,

$\Rightarrow$ Factor by groups considering the combined terms have a common term.

Example

$$
\begin{aligned}
x^3 + 3x^2 - 4x - 12 &= (x^3 + 3x^2) + (-4x - 12) \quad \text{Grouping} \\
&= x^2(x + 3) - 4(x + 3) \\
&= (x + 3)(x^2 - 4) \quad \text{Find GCF} \\
&= (x + 3)(x + 2)(x - 2) \quad \text{Factorization}
\end{aligned}
$$

The GCF of the first group $(x^3 + 3x^2)$ is x^2.

Factor it out of the expression to get $x^2(x + 3)$.

The GCF of the second group $(-4x - 12)$ is -4.

Factor it out of the expression to get $-4(x + 3)$.

Consider the same coefficient a in the form of $x^2 + ax + b$.

$$(x + 1)(x + 2)(x + 3)(x + 4) - 8$$
$$= (x + 1)(x + 4)\ (x + 2)(x + 3) - 8 \quad \text{Grouping}$$
$$= (x^2 + 5x + 4)(x^2 + 5x + 6) - 8$$
$$= (A + 4)(A + 6) - 8 \quad \text{Substitute } x^2 + 5x \text{ as } A$$
$$= A^2 + 10A + 24 - 8$$
$$= A^2 + 10A + 16$$
$$= (A + 2)(A + 8)$$
$$= (x^2 + 5x + 2)(x^2 + 5x + 8) \quad \text{Replace A with } x^2 + 5x$$

4. If polynomials contain more than 2 variables,

⇒ Rewrite the polynomials in the descending order of powers for the variable with the lowest power.

Consider the other terms (which don't contain the variable) as constants.

Then factor the pretended constant term.

> x has a power of 2.
>
> y has a power of 1.
>
> Because y has the lowest power, choose it for the descending order of powers.

Example

$$x^2 + 2x + xy - y - 3 = (xy - y) + \underline{(x^2 + 2x - 3)} \qquad \text{Descending order of powers of } y$$

$$ \text{L}\text{-----} \text{Pretended constant term}$$

$$= y(x - 1) + (x^2 + 2x - 3) \qquad \text{Factor a group}$$

$$= y(x - 1) + (x + 3)(x - 1) \qquad \text{Factor a group}$$

$$= (x - 1)(y + x + 3) \qquad \text{Common factor}$$

$$= (x - 1)(x + y + 3)$$

Exercises

#1. Find all factors for the following expressions:

  (1) $ab + a + b + 1$

  (2) $abx + aby$

  (3) $2x^2 + xy$

#2. Factor the following polynomials:

  (1) $a^2 - ab + a$

  (2) $a^2b - ab^2$

  (3) $4a - 10$

  (4) $2a^3 + 3a^2 - 5a$

  (5) $9a^4b^2 - 3a^3b^2 + 12a^2b^2$

  (6) $2a(x + y) - 4b(x + y)$

  (7) $-12(x - 2y) - 18a(x - 2y)$

  (8) $4a(a - b) + 4b(b - a)$

  (9) $a^n + a^{n+2}$

  (10) $3x^{2n} + 12x^{3n} + 9x^n$

#3. Factor each polynomial using factorization formulas.

  (1) $x^2 - 2x + 1$

  (2) $9x^2 + 6x + 1$

  (3) $4x^2 - 4x + 1$

  (4) $x^2 + 10x + 25$

  (5) $x^2 + x + \frac{1}{4}$

  (6) $x^6 + 6x^3 + 9$

  (7) $x^{2n} - 2x^n y^n + y^{2n}$

  (8) $x^2 - 1$

  (9) $x^2 - 4y^2$

  (10) $9x^2 - \frac{1}{4}y^2$

  (11) $(x + a)^2 - 36$

  (12) $1 - (x + y)^2$

  (13) $x^4 - 1$

  (14) $x^8 - y^8$

  (15) $-16x^2 + 9y^2$

  (16) $\frac{1}{9}x^2 - \frac{9}{16}y^2$

  (17) $4x^3y - xy^3$

  (18) $x^2 + 4x + 3$

  (19) $x^2 - 4x - 5$

  (20) $x^2 - 3x + 2$

  (21) $x^2 - 2x - 8$

  (22) $(x - y)^2 - (x + y)^2$

  (23) $(2x - 3)^2 - (x + 1)^2$

  (24) $x^2 - \frac{5}{6}x + \frac{1}{6}$

  (25) $3x^2 - 4x - 4$

  (26) $2x^2 + 3x - 2$

  (27) $4x^2 - 2x - 12$

  (28) $4x^2 - 10x - 6$

  (29) $2x^2 - 3xy - 9y^2$

  (30) $x^3y - 16xy^3$

(31) $\frac{1}{3}x^2 - 2 + \frac{3}{x^2}$

(32) $4x^2 - 12xy + 9y^2$

(33) $\frac{1}{3}x^2 - \frac{1}{3}x - 2$

(34) $a^4(x - y) + b^4(y - x)$

(35) $-3a^2 + 3a + 6$

(36) $3ax^2 - 5ax - 2a$

(37) $(a - 1)^2 - 10(a - 1) + 25$

(38) $a^8 - 1$

(39) $(x - 1)(x + 2) - 4$

(40) $4x^2 - 2x - 2 - (x - 1)^2$

#4. Find the value of k that will make the following polynomials perfect square forms.

(1) $x^2 + 5x + k$

(2) $9x^2 - 12x + k$

(3) $2x^2 - 6x + k$

(4) $25x^2 + 4x + k$

(5) $\frac{1}{25}x^2 + kx + 4$

(6) $4x^2 - kx + 25$

(7) $2x^2 + kx + 8$

(8) $9x^2 + (2k - 4)x + 4$

(9) $4x^2 + (k + 5)x + 9y^2$

(10) $k - \frac{1}{4}xy + \frac{1}{4}y^2$

(11) $9x^2 + (k - 1)xy + 25y^2$

(12) $kx^2 + 3x + 9$

(13) $(x + 3)(x - 4) - k$

(14) $(2x + 1)(2x - 4) + k$

(15) $(x - 1)(x - 2)(x + 4)(x + 5) + k$

#5. Find the value of a for the following polynomials. Each polynomial has a given factor.

(1) $x^2 + 2x + a$ has the factor $(x + 3)$.

(2) $3x^2 + ax - 8$ has the factor $(x - 2)$.

(3) $4x^2 + ax - 6$ has the factor $(3 - 2x)$.

(4) $2x^2 + (3a - 1)x - 15$ has the factor $(2x + 3)$.

(5) $2ax^2 - 5x + 2$ has the factor $(3x - 2)$.

#6. Find the value of $a + b$ for any constants a and b.

(1) $3ax^2 - 6x + ab$ has the factor $(3x - 1)^2$.

(2) $ax^2 + 8x + 4b$ has two factors $(3x + 2)$ and $(x - 2)$.

(3) $(4x - 3)^2 - (3x - 2)^2$ has two factors $(ax + 5)$ and $(b - x)$.

(4) $2x^2 + ax - 4$ and $bx^2 - x - 2$ have the same factor $(2x + 1)$.

#7. For any integers $a(\neq 0)$ and b, the length and width of a rectangle are forms of $ax + b$.

Find the perimeter of a rectangle whose area is $6x^2 + 17x + 12$.

#8. Find a possible expression for the width of a rectangle whose area is $12x^2 + 5x - a$,

where $a > 0$ and the width is greater than the length.

#9. The area and base of a triangle are $5x^2 + 12x + 4$ and $2x + 4$, respectively.

Find the height of the triangle.

#10. The figures $A, B,$ and C (equilateral) have the same area. Find the perimeters of A and C.

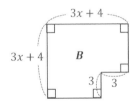
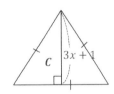

#11. A and B are squares with the lengths a and b, respectively. The perimeter of A is 40 more

than the perimeter of B. The difference between their areas is 200. Find the sum of their areas.

#12. Factor each polynomial using any method.

(1) $3a^2b - a^3b - 2ab$

(2) $3a - 3a^2 + 6$

(3) $8a^2b + 6ab - 20b$

(4) $a^5 - 16a$

(5) $a^2(x - y) + b^2(y - x)$

(6) $ab(a - b) + 2a(b - a)^2$

(7) $(a - b)^2 - 2(b - a)^3$

(8) $a(x - y)^2 + b(x - y)$

(9) $4(x + 2)^2 - 3(x + 2) - 10$

(10) $(x - y)(x - y + 3) - 4$

(11) $(2x - y)^2 + 8y(2x - y) + 16y^2$

(12) $2(x - 1)^2 - 3(x - 1)(y + 1)$
$\qquad -9(y + 1)^2$

(13) $a^4 - 5a^2 - 36$

(14) $a^8 - 2a^4 - 8$

(15) $x^2 - xy + 2x - 2y$

(16) $a^2 - ab - b - 1$

(17) $a^3 - a^2 - a + 1$

(18) $a^4 + 3a - 3a^3 - a^2$

(19) $2ab + 1 - a^2 - b^2$

(20) $(x + 2)(x - 1)^2(x - 4) - 10$

(21) $9x^2 - y^2 - 4y - 4$

(22) $a^2 - b^2 + 4b - 4$

(23) $a^2 - 16b^2 - 6a + 9$

(24) $ax^2 - a - bx^2 + b$

(25) $x^2 - xy + x + 2y - 6$

(26) $-2a^2 - 5a + 2ab - b + 3$

(27) $2x^2 - y^2 + xy - 2x + y$

(28) $4a^2 - b^2 - 4a + 4b - 3$

(29) $4a^2 - 4ab + b^2 - c^2$

(30) $a^4 + a^2 + 1$

(31) $a^4 - 6a^2 + 1$

(32) $a^4 - 13a^2 + 4$

(33) $9x^4 + 8x^2 + 4$

#13. Evaluate each expression using factorization.

(1) $99^2 - 1$

(2) $99^2 - 89^2$

(3) $49^2 - 51^2$

(4) $3^8 - 1$

(5) $6^2 - 5^2 + 4^2 - 3^2 + 2^2 - 1$

(6) $\left(1 - \dfrac{1}{2^2}\right)\left(1 - \dfrac{1}{3^2}\right)\left(1 - \dfrac{1}{4^2}\right)\cdots$

$\cdots \left(1 - \dfrac{1}{99^2}\right)\left(1 - \dfrac{1}{100^2}\right)$

(7) $3(2^2 + 1)(2^4 + 1)(2^8 + 1) + 1$

(8) $\dfrac{99 \times 101 + 99 \times 2}{101^2 - 4}$

(9) $36 \times 34 - 35 \times 34$

(10) $87 \times 56 + 87 \times 44$

(11) $65^2 - 2 \times 65 \times 35 + 35^2$

(12) $25^2 + 30 \times 25 + 15^2$

(13) $a^2 - 8a - 20$ when $a = 28$

(14) $a^2 + 3a - 54$ when $a = 91$

Chapter 13. Quadratic Equations

13-1 Quadratic Equations and Their Solutions

1. Definition

(1) A *quadratic equation* which has transferred all the terms on the right side of the equal sign to the left side of the equal sign is formed by

$$ax^2 + bx + c = 0, \text{ for any constants } a \neq 0, \; b, \text{ and } c.$$

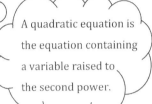

A quadratic equation is the equation containing a variable raised to the second power.

Note : Quadratic equations : $ax^2 + bx + c = 0, \; a \neq 0$

$ax^2 + bx = 0, \; a \neq 0 \; (\text{When } c = 0)$

$ax^2 + c = 0, \; a \neq 0 \; (\text{When } b = 0)$

$ax^2 = 0, \; a \neq 0 \; (\text{When } b = 0 \text{ and } c = 0)$

(2) The *solution* (*root*) of an unknown variable, x,

must make the quadratic equation $(ax^2 + bx + c = 0, a \neq 0)$ always true.

Whether the equation $ax^2 + bx + c = 0, a \neq 0$ is true or not depends on the value of the variable x.

Example $x^2 - 2x + 1 = 0$

If $x = 1$, then $1 - 2 + 1 = 0$ (this makes the quadratic equation true.)

∴ $x = 1$ is a solution.

If $x = -1$, then $1 + 2 + 1 \neq 0$ (this makes the quadratic equation untrue.)

∴ $x = -1$ is not a solution.

A quadratic equation $ax^2 + bx + c = 0, a \neq 0$ has up to two different real number solutions.

2. Solving Quadratic Equations

(1) Using Factorization and The Zero Product Property

1) The Zero product property

For any expressions or numbers A and B,

$$AB = 0 \underset{\text{if and only if}}{\Longleftrightarrow} A = 0 \text{ or } B = 0$$

To apply the zero product property,

if $ax^2 + bx + c = k$,

$\Rightarrow ax^2 + bx + c - k = 0$

(∵ One side of equal sign must equal 0.)

Note : $A = 0$ or $B = 0 \Leftrightarrow$ $\begin{cases} A = 0 \ and \ B \neq 0 \\ A \neq 0 \ and \ B = 0 \\ A = 0 \ and \ B = 0 \end{cases}$

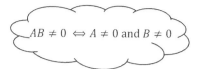

$AB \neq 0 \Leftrightarrow A \neq 0$ and $B \neq 0$

2) Steps for Solving Quadratic Equations

Step 1: Simplify the equation as a form of $ax^2 + bx + c = 0, a \neq 0$

Step 2: Factor the left side of the equation into the form of $a(x - \alpha)(x - \beta) = 0$

Step 3: Solve the equation using the zero product property;

$$(x - \alpha) = 0 \ \text{ or } \ (x - \beta) = 0 \ ; \ x = \alpha \ \text{ or } \ x = \beta$$

Example $\quad 2x^2 - 4x - 6 = 0$

$\underset{\text{Step 1}}{\Longrightarrow} \quad 2(x^2 - 2x - 3) = 0$

$\underset{\text{Step 2}}{\Longrightarrow} \quad 2(x - 3)(x + 1) = 0$

$\underset{\text{Step 3}}{\Longrightarrow} \quad x = 3 \ \text{ or } \ x = -1$

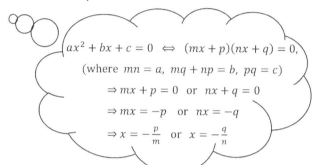

$ax^2 + bx + c = 0 \Leftrightarrow (mx + p)(nx + q) = 0,$

(where $mn = a, \ mq + np = b, \ pq = c$)

$\Rightarrow mx + p = 0 \ \text{ or } \ nx + q = 0$

$\Rightarrow mx = -p \ \text{ or } \ nx = -q$

$\Rightarrow x = -\dfrac{p}{m} \ \text{ or } \ x = -\dfrac{q}{n}$

3) Double roots

If a quadratic equation has a same factor twice, then the equation has a double root.

$ax^2 + bx + c = 0 \ \Rightarrow \ a(x - \underline{\alpha})(x - \underline{\alpha}) = 0$ These are two identical factors.

$\Rightarrow a(x - \alpha)^2 = 0$

$\Rightarrow x - \alpha = 0 \ (\because a \neq 0)$

$\Rightarrow x = \alpha$ (The double root)

$a(x - \alpha)^2 = 0$

$\Rightarrow a(x - \alpha)(x - \alpha) = 0$

$\Rightarrow (x - \alpha)(x - \alpha) = 0 \ (\because a \neq 0)$

$\Rightarrow x = \alpha$ or $x = \alpha \ (\because$ by the zero product property)

$\Rightarrow x = \alpha$: the double root (when you get the same answer twice)

Example

$x^2 + 4x + 4 = 0 \ \Rightarrow \ (x + 2)^2 = 0 \ \Rightarrow \ x = -2$ (The double root)

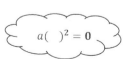

$a(\ \ )^2 = 0$

Example

If the equation $x^2 - 3x + k = 0$ has a double root, $k = \left(\dfrac{-3}{2}\right)^2 = \dfrac{9}{4}$.

So, $x^2 - 3x + k = x^2 - 3x + \dfrac{9}{4} = \left(x - \dfrac{3}{2}\right)^2 = 0$

$\therefore \; x = \dfrac{3}{2}$ (The double root)

$x^2 + ax + \left(\dfrac{a}{2}\right)^2 = \left(x + \dfrac{a}{2}\right)^2 = 0$

Constant $= \left(\dfrac{x-\text{coefficient}}{2}\right)^2$, when x^2-coefficient $= 1$

① $a(x - \alpha)^2 = 0, \; a \neq 0 \; \Rightarrow \; x = \alpha$ (The double root)

② To find a double root for the quadratic equation $a(x - p)^2 = q$,

$q = 0$ (q must equal 0). So, we get $x = p$ (The double root)

Note: ① *If the quadratic equation $x^2 = q$ has solutions, $q \geq 0$ (q must be positive or zero).*

② *If the quadratic equation $x^2 = q$ has 2 different solutions, $q > 0$ (q must be positive).*

③ *If the quadratic equation $x^2 = q$ has a double root, $q = 0$ (q must be zero).*

(2) Using Perfect Squares (Completing the Square)

Solve the equation in the form of a perfect square: $(x + p)^2 = q$

1) Square roots

If the quadratic equation $ax^2 + bx + c = 0$ is formed by $(x + p)^2 = q$,

$\Rightarrow$ The solution is

$$\begin{cases} x = -p \pm \sqrt{q} & , \; q > 0 \\ x = -p \text{ (A double root)} & , \; q = 0 \\ \text{No real solution} & , \; q < 0 \end{cases}$$

If the real solution for the quadratic equation $x^2 = q$

does not exist (No solution), then $q < 0$ (q must be negative).

Note: ① $x^2 = p, \; p \geq 0 \; \Rightarrow x = \pm\sqrt{p}$

② $ax^2 = p, \; a \neq 0, \; \dfrac{p}{a} \geq 0 \; \Rightarrow x^2 = \dfrac{p}{a} \; \Rightarrow \; x = \pm\sqrt{\dfrac{p}{a}}$

③ $(x + p)^2 = q, \; q \geq 0 \; \Rightarrow x + p = \pm\sqrt{q} \; \Rightarrow \; x = -p \pm \sqrt{q}$

④ $(ax + p)^2 = q, \; a \neq 0, \; q \geq 0 \; \Rightarrow ax + p = \pm\sqrt{q} \; \Rightarrow \; x = \dfrac{-p \pm \sqrt{q}}{a}$

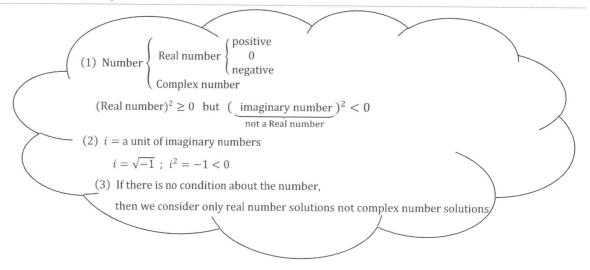

(1) Number $\begin{cases} \text{Real number} \begin{cases} \text{positive} \\ 0 \\ \text{negative} \end{cases} \\ \text{Complex number} \end{cases}$

(Real number)$^2 \geq 0$ but $(\underbrace{\text{imaginary number}}_{\text{not a Real number}})^2 < 0$

(2) i = a unit of imaginary numbers

$i = \sqrt{-1}$; $i^2 = -1 < 0$

(3) If there is no condition about the number,

then we consider only real number solutions not complex number solutions

2) Steps for Solving Quadratic Equations Using Perfect Squares

Step 1. Make the coefficient of x^2 equal to 1.

$$ax^2 + bx + c = 0 \quad \Rightarrow \quad a\left(x^2 + \frac{b}{a}x + \frac{c}{a}\right) = 0$$

Step 2. Divide by a on both sides or use the zero product property.

$$\Rightarrow \quad x^2 + \frac{b}{a}x + \frac{c}{a} = 0$$

Step 3. Transfer the constant to the right side of the equal sign by subtracting $\frac{c}{a}$ from both sides.

$$\Rightarrow \quad x^2 + \frac{b}{a}x = -\frac{c}{a}$$

Step 4. Square of half the coefficient of x. Add the result to both sides of the equal sign.

$$x^2 + \frac{b}{a}x + \left(\frac{b}{2a}\right)^2 = -\frac{c}{a} + \left(\frac{b}{2a}\right)^2 \quad \text{Completing the square}$$

$$\Rightarrow \left(x + \frac{b}{2a}\right)^2 = -\frac{c}{a} + \left(\frac{b}{2a}\right)^2 \qquad \text{Form of perfect square, } (x+p)^2 = q$$

OR

$$x^2 + \frac{b}{a}x = \left(x + \frac{1}{2}\cdot\frac{b}{a}\right)^2 - \left(\frac{b}{2a}\right)^2 = -\frac{c}{a}$$

$$\Rightarrow \left(x + \frac{1}{2}\cdot\frac{b}{a}\right)^2 = -\frac{c}{a} + \left(\frac{b}{2a}\right)^2 \qquad \text{Form of perfect square, } (x+p)^2 = q$$

Step 5. Solve the equation using square root.

$$x^2 + mx = \left(x + \frac{1}{2}m\right)^2 - \left(\frac{1}{2}m\right)^2$$

$\dfrac{1}{2}\cdot\dfrac{b}{a} = \dfrac{1}{2}\cdot \text{(the coefficient of } x)$

Since $(x \pm m)^2 = x^2 \pm 2mx + m^2$,

$$\left(x + \frac{1}{2}m\right)^2 = x^2 + 2\cdot\frac{1}{2}mx + \left(\frac{1}{2}m\right)^2 = x^2 + mx + \left(\frac{1}{2}m\right)^2$$

$$\therefore \ x^2 + mx = \left(x + \frac{1}{2}m\right)^2 - \left(\frac{1}{2}m\right)^2$$

Example

$$2x^2 + 5x + 3 = 0$$

$$\Longrightarrow \quad 2\left(x^2 + \frac{5}{2}x + \frac{3}{2}\right) = 0$$

$$\underset{\text{divide by 2}}{\Longrightarrow} \quad x^2 + \frac{5}{2}x + \frac{3}{2} = 0$$

$$\underset{\text{transfer constant}}{\Longrightarrow} \quad x^2 + \frac{5}{2}x = -\frac{3}{2} \quad \text{Move the constant to the right side of the equation}$$

$$\Longrightarrow \quad \left(x + \frac{1}{2}\cdot\frac{5}{2}\right)^2 - \left(\frac{1}{2}\cdot\frac{5}{2}\right)^2 = -\frac{3}{2} \quad \text{To find the perfect square}$$

$$\Longrightarrow \quad \left(x + \frac{5}{4}\right)^2 = -\frac{3}{2} + \left(\frac{5}{4}\right)^2$$

$$\Longrightarrow \quad \left(x + \frac{5}{4}\right)^2 = \frac{1}{16} \quad \text{Perfect square form}$$

$$\underset{\text{use the square root}}{\Longrightarrow} \quad x + \frac{5}{4} = \pm\sqrt{\frac{1}{16}} \quad \text{To solve the equation for } x$$

$$\Longrightarrow \quad x = -\frac{5}{4} \pm \frac{1}{4}$$

$$\Longrightarrow \quad x = -1 \ \text{ or } \ -\frac{6}{4}\left(= -\frac{3}{2}\right) \quad \text{The solution to the equation}$$

(3) Using Quadratic Formulas

If an equation is not factorized, solve it using quadratic formulas.

Before using a quadratic formula to a quadratic equation, you must have the equation in standard form, $ax^2 + bx + c = 0, \ a \neq 0$.

1) Quadratic Formula I

$$ax^2 + bx + c = 0, \ a \neq 0$$

$$\Rightarrow \quad x = \frac{-b \pm \sqrt{b^2 - 4ac}}{2a}, \ b^2 - 4ac \geq 0$$

2) Quadratic Formula II (When $b = 2b'$ is an even number)

$$ax^2 + 2b'x + c = 0, \ a \neq 0$$

$$\Rightarrow \quad x = \frac{-b' \pm \sqrt{b'^2 - ac}}{a}, \ b'^2 - ac \geq 0$$

When the coefficient of x is an even number, the Quadratic Formula II can be easily used.

Note $ax^2 + bx + c = 0,\ a \neq 0$ *Standard form of general equation*

$$\Rightarrow\ a\left(x^2 + \frac{b}{a}x + \frac{c}{a}\right) = 0$$

$$\Rightarrow\ x^2 + \frac{b}{a}x + \frac{c}{a} = 0 \qquad \textit{Divide each side by } a$$

$$\Rightarrow\ x^2 + \frac{b}{a}x = -\frac{c}{a} \qquad \textit{Subtract } -\frac{c}{a} \textit{ from each side}$$

$$\Rightarrow\ \left(x + \frac{1}{2}\cdot\frac{b}{a}\right)^2 - \left(\frac{b}{2a}\right)^2 = -\frac{c}{a} \qquad \left(\frac{b}{2a}\right)^2 \textit{The square of half the coefficient of } x$$

$$\Rightarrow\ \left(x + \frac{b}{2a}\right)^2 = \frac{b^2 - 4ac}{4a^2} \qquad \textit{Add } \left(\frac{b}{2a}\right)^2 \textit{ to each side}$$

$$\Rightarrow\ x + \frac{b}{2a} = \pm\sqrt{\frac{b^2 - 4ac}{4a^2}} \qquad \textit{Take square roots of each side}$$

$$\Rightarrow\ x = -\frac{b}{2a} \pm \sqrt{\frac{b^2 - 4ac}{4a^2}} \qquad \textit{Subtract } -\frac{b}{2a} \textit{ from each side}$$

$$\Rightarrow\ x = \frac{-b \pm \sqrt{b^2 - 4ac}}{2a} \qquad \textit{Simplify}$$

Example

$x^2 + 5x + 3 = 0$ (This equation can't be factorized.)

$$\xrightarrow[a=1,b=5,c=3]{}\quad x = \frac{-5 \pm \sqrt{(5)^2 - 4\cdot1\cdot3}}{2\cdot1} = \frac{-5 \pm \sqrt{13}}{2} \qquad \text{Using Quadratic Formula I}$$

$$\therefore\ x = \frac{-5 + \sqrt{13}}{2} \ \text{ or } \ x = \frac{-5 - \sqrt{13}}{2}$$

Example

$4x^2 + 4x - 3 = 0$

① Using Quadratic Formula I

$$\xrightarrow[a=4,b=4,c=-3]{}\quad x = \frac{-4 \pm \sqrt{(4)^2 - 4\cdot4\cdot(-3)}}{2\cdot4} = \frac{-4 \pm \sqrt{64}}{8} = \frac{-4 \pm 8}{8}$$

$$\therefore\ x = \frac{1}{2} \ \text{ or } \ x = -\frac{3}{2}$$

② Using Quadratic Formula II

$$\xrightarrow[a=4,b\prime=2,c=-3]{}\quad x = \frac{-2 \pm \sqrt{(2)^2 - 4\cdot(-3)}}{4} = \frac{-2 \pm \sqrt{16}}{4} = \frac{-2 \pm 4}{4}$$

$$\therefore\ x = \frac{1}{2} \ \text{ or } \ x = -\frac{3}{2}$$

③ Using Factorization

$$4x^2 + 4x - 3 = (2x - 1)(2x + 3) = 0$$

$$\therefore\ x = \frac{1}{2} \ \text{ or } \ x = -\frac{3}{2}$$

④ Using a Perfect Square

$$4x^2 + 4x - 3 = 0 \Rightarrow 4\left(x^2 + x - \frac{3}{4}\right) = 0 \Rightarrow x^2 + x - \frac{3}{4} = 0$$

$$\Rightarrow \left(x + \frac{1}{2}\right)^2 - \frac{1}{4} = \frac{3}{4} \Rightarrow \left(x + \frac{1}{2}\right)^2 = 1$$

$$\Rightarrow x + \frac{1}{2} = \pm 1 \Rightarrow x = -\frac{1}{2} \pm 1$$

$$\therefore x = \frac{1}{2} \text{ or } x = -\frac{3}{2}$$

> The solutions of a quadratic equation depend on the sign of the number D under the symbol $\sqrt{\ }$.
> In order to have possible solutions, $D \geq 0$.

3. The Discriminant $D = b^2 - 4ac$

For a quadratic equation of the form $ax^2 + bx + c = 0$, the solution is

$$x = \frac{-b \pm \sqrt{b^2 - 4ac}}{2a}, \quad b^2 - 4ac \geq 0$$

In the quadratic formula, the expression $b^2 - 4ac$ (denoted by D), underneath the radical sign, is called the *Discriminant* of the quadratic equation $ax^2 + bx + c = 0$ and the discriminant determines the number and the different types of solutions for the equation.

---

① If $D > 0$, then the equation has two distinct roots

$$x = \frac{-b + \sqrt{b^2 - 4ac}}{2a} \quad \text{and} \quad x = \frac{-b - \sqrt{b^2 - 4ac}}{2a} \quad \text{(2 different real number solutions)}$$

② If $D = 0$, then the equation has a double root

$$x = \frac{-b}{2a} \quad \text{(Only 1 real number solution; the same real number solution twice)}$$

③ If $D < 0$, then the equation has two distinct complex number solutions

(No real number solution)

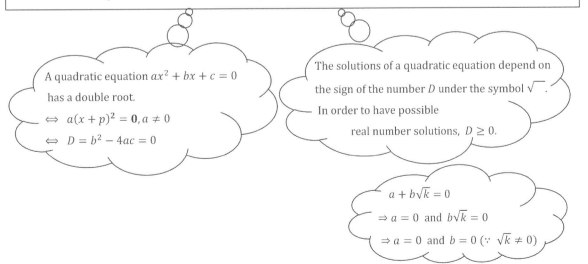

> A quadratic equation $ax^2 + bx + c = 0$ has a double root.
> $\Leftrightarrow a(x + p)^2 = 0, a \neq 0$
> $\Leftrightarrow D = b^2 - 4ac = 0$

> The solutions of a quadratic equation depend on the sign of the number D under the symbol $\sqrt{\ }$.
> In order to have possible real number solutions, $D \geq 0$.

> $a + b\sqrt{k} = 0$
> $\Rightarrow a = 0$ and $b\sqrt{k} = 0$
> $\Rightarrow a = 0$ and $b = 0$ ($\because \sqrt{k} \neq 0$)

4. The Solution-Coefficient Relationship

The solutions and the coefficients of a quadratic equation are related to the obtaining the sum and the product of the solutions. We can also obtain a quadratic equation from the solutions.

(1) Given a quadratic equation $ax^2 + bx + c = 0$, suppose the solutions are $x = \alpha$ and $x = \beta$. Then we obtain:

> If $\alpha\beta = \frac{c}{a} < 0 \Rightarrow ac < 0$
>
> $\Rightarrow D = b^2 - 4ac > 0 \ (\because b^2 > 0 \text{ and } -(ac) > 0)$
>
> $\Rightarrow$ 2 different solutions exist.

$$\boxed{\begin{aligned} &1) \ \alpha + \beta = -\frac{b}{a} \quad \text{(The sum of the solutions)} \\ &2) \ \alpha \cdot \beta = \frac{c}{a} \quad \text{(The product of the solutions)} \end{aligned}}$$

Note : $ax^2 + bx + c = 0, \ a \neq 0$

$\Rightarrow \ x = \dfrac{-b+\sqrt{b^2-4ac}}{2a} \ or \ x = \dfrac{-b-\sqrt{b^2-4ac}}{2a}$

> $(\alpha \pm \beta)^2 = \alpha^2 \pm 2\alpha\beta + \beta^2$
> $\alpha^2 + \beta^2 = (\alpha+\beta)^2 - 2\alpha\beta$
> $(\alpha+\beta)^2 = (\alpha-\beta)^2 + 4\alpha\beta$
> $(\alpha-\beta)^2 = (\alpha+\beta)^2 - 4\alpha\beta$
> $\alpha^2 - \beta^2 = (\alpha+\beta)(\alpha-\beta)$

Let $\ \alpha = \dfrac{-b+\sqrt{b^2-4ac}}{2a} \ and \ \beta = \dfrac{-b-\sqrt{b^2-4ac}}{2a}$

Then, $\ \alpha + \beta = \dfrac{-b+\sqrt{b^2-4ac}}{2a} + \dfrac{-b-\sqrt{b^2-4ac}}{2a} = \dfrac{-2b}{2a} = -\dfrac{b}{a}$

$\alpha \cdot \beta = \dfrac{-b+\sqrt{b^2-4ac}}{2a} \cdot \dfrac{-b-\sqrt{b^2-4ac}}{2a} = \dfrac{(-b)^2-\left(\sqrt{b^2-4ac}\right)^2}{4a^2} = \dfrac{b^2-(b^2-4ac)}{4a^2} = \dfrac{4ac}{4a^2} = \dfrac{c}{a}$

(2) If the solutions are given, then a quadratic equation can be obtained.

1) $\begin{pmatrix} x^2\text{-coefficient} =a \ (\neq 0) \\ \text{Solutions} : x= \alpha \ and \ x=\beta \end{pmatrix} \Rightarrow$ The quadratic equation is $a(x- \alpha)(x- \beta)=0$

2) $\begin{pmatrix} x^2\text{-coefficient} =a \ (\neq 0) \\ \text{Solution} : x= \alpha \ (\text{the double root}) \end{pmatrix} \Rightarrow$ The quadratic equation is $a(x- \alpha)^2=0$

3) $\begin{pmatrix} x^2\text{-coefficient} =a \ (\neq 0) \\ \text{The sum of the solutions}=p, \\ \text{The product of the solutions}=q \end{pmatrix} \Rightarrow$ The quadratic equation is $a(x^2- px + q)=0$

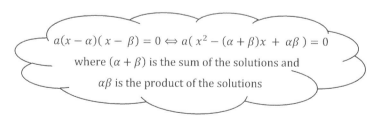

> $a(x - \alpha)(x - \beta) = 0 \Leftrightarrow a(x^2 - (\alpha + \beta)x + \alpha\beta) = 0$
> where $(\alpha + \beta)$ is the sum of the solutions and $\alpha\beta$ is the product of the solutions

(3) If a quadratic equation has a solution $\alpha + \beta\sqrt{m}$, then the other solution is $\alpha - \beta\sqrt{m}$.

Note $ax^2 + bx + c = 0, a \neq 0$

$\Rightarrow \quad x = \dfrac{-b + \sqrt{b^2 - 4ac}}{2a}$ (*A form of* $\alpha + \beta \sqrt{m}$)

$or \quad x = \dfrac{-b - \sqrt{b^2 - 4ac}}{2a}$ (*A form of* $\alpha - \beta \sqrt{m}$)

Example

With given solutions, we can find a quadratic equation.

If the x^2-coefficient is 3 and the solutions are $x = 1$ and $x = 2$, then the quadratic equation is

$3(x - 1)(x - 2) = 0$; $3(x^2 - 3x + 2) = 0$ or

$3(x^2 - (1 + 2)x + (1 \cdot 2)) = 0$; $3(x^2 - 3x + 2) = 0$

Example

With the coefficients of a quadratic equation, we can find the solutions.

If the quadratic equation $x^2 - 7x + 12 = 0$ has solutions $x = \alpha$ and $x = \beta$, then

$\alpha + \beta = -\dfrac{-7}{1} = 7$ and $\alpha \cdot \beta = \dfrac{12}{1} = 12$

$\therefore \ (\alpha - \beta)^2 = (\alpha + \beta)^2 - 4\alpha\beta = 7^2 - 4 \cdot 12 = 1$; $\alpha - \beta = \pm 1$

$\therefore \qquad \alpha + \beta = 7$

$\underline{+) \ \alpha - \beta = \pm 1}$

$\quad 2\alpha \quad = 7 \pm 1$; $\alpha = 4$ or $\alpha = 3$

Therefore, the solutions are $x = 4$ and $x = 3$.

Exercises

#1. State whether each expression is a quadratic equation (Yes) or is not a quadratic equation (No).

(1) $x^2 = 5$

(2) $x(x + 3) = 0$

(3) $x^2 = (x + 3)^2$

(4) $x^3 + x^2 + 1 = 0$

(5) $2x^3 + x^2 = 3x + 2x^3 + 5$

(6) $\frac{1}{x^2} + \frac{1}{x} + 3 = 0$

(7) $(x + 1)^2 - (x - 1)^2 + 3 = 0$·

(8) $2x^2 + 5x = (x + 1)(x + 2)$

(9) $x^2 + 4x + 4 = 2(x + 2)^2 - 2$

(10) $x^2 - 5x + 6$

(11) $3x^2 + 5x + 1 = 2x^2 - 6x + 4$

(12) $\frac{x^2}{2} + 4x = 4x + 4$

(13) $\frac{2}{x^2} + 2x = 3$

(14) $x^2 + 1 = (x + 1)^2$

(15) $x^2 + 2x + 1 = 2(x + 1)^2$

(16) $x^2 = 0$

(17) $x(2x + 1) = 3x(x + 2)$

(18) $(x + 4)^2$

(19) $x^2(x - 1) = x(x^2 + x - 1)$

(20) $x^2 + 3x = x^2 + 3$

#2. The following equations are quadratic equations. Find the condition for constants a and b.

(1) $(x + 1)(ax + 2) = 2x^2 + 5$

(2) $2(x^2 + 2x + 1) = (x + 2)(5 - ax)$

(3) $(3x - 1)(ax + 2) = 5 - bx^2$

(4) $(2x + 1)(3x + 2) = (ax + 2)(bx - 3)$

(5) $(2a + b)x^2 + ax + b = 0$

(6) $a^2x^2 + bx + 5 = 5$

#3. Find the value of $a + b + c$ for the quadratic equation $ax^2 + bx + c = 0$, in which a is the smallest positive number.

(1) $2(x - 1)^2 = (x + 1)^2 + 5$

(2) $(3x + 1)(x - 2) = 2 - x^2$

(3) $3(x + 1)^2 = 3(x + 1)$

(4) $x^2 = x$

(5) $2x(x - 1) = x^2 - 2$

#4. Find the sum of the solutions for each quadratic equation using factorization.

(1) $x^2 - 2x - 3 = 0$

(2) $2x^2 - 7x + 5 = 0$

(3) $-3x^2 + 6x = 0$

(4) $2x^2 + 2x - 4 = 0$

(5) $x(x + 5) = 6$

(6) $x^2 = \frac{x+1}{2}$

#5. Each of the following quadratic equations has a solution $x = \alpha$. For each equation, find the value of constant a and the other solution $x = \beta$ for the equation.

(1) $x^2 + ax - 4 = 0$, $x = 1$

(2) $3x^2 - ax + a = 0$, $x = -1$

(3) $2x^2 - x + a = 0$, $x = -1$

(4) $ax^2 - 2x - 3 = 0$, $x = -2$

(5) $ax^2 - (a - 1)x - 6 = 0$, $x = 2$

#6. Each of the following quadratic equations has the solution $x = \alpha$.

Find the value of the given expression for each equation.

(1) $\alpha - \dfrac{1}{\alpha}$ for $x^2 - 2x + 1 = 0$

(2) $\alpha^2 + \dfrac{1}{\alpha^2}$ for $2x^2 + 3x - 2 = 0$

(3) $\left(\alpha + \dfrac{1}{\alpha}\right)^2$ for $x^2 - 3x - 1 = 0$

(4) $\alpha^2 + \dfrac{9}{\alpha^2}$ for $3x^2 + 2x - 9 = 0$

(5) $\dfrac{\alpha-1}{\alpha+1} - \dfrac{\alpha+1}{\alpha-1}$ for $2x^2 - 6x - 2 = 0$

(6) $\alpha^2 + \alpha - \dfrac{1}{\alpha} + \dfrac{1}{\alpha^2}$ for $x^2 - 4x - 1 = 0$

#7. Find the value of the given expression for the following quadratic equations.

Each equation has two solutions (α, β) where $\alpha > \beta$.

(1) $\alpha + \beta$ for $x^2 - 5x - 6 = 0$

(2) $\alpha - \beta$ for $x^2 + 3x - 10 = 0$

(3) $\dfrac{\alpha+\beta}{\alpha-\beta}$ for $12x^2 + 5x - 3 = 0$

(4) $\alpha^2 - \beta^2$ for $6x^2 - x - 1 = 0$

#8. Solve the following quadratic equations using square roots

(1) $2x^2 = 8$

(2) $9x^2 - 5 = 0$

(3) $3(x - 1)^2 = 15$

(4) $(2x + 5)^2 - 3 = 0$

(5) $4(x - 2)^2 - 1 = 0$

#9. Solve the following quadratic equations using perfect squares

(1) $x^2 - 3x - 3 = 0$

(2) $2x^2 + 5x = 7$

(3) $-x^2 - 3x + 5 = 0$

(4) $3x^2 - 4x + 1 = 0$

#10. Find the constant k for the following quadratic equations with a double root

(1) $(3x - 4)^2 - k^2 = 0$

(2) $x^2 - kx + 5 = 0$

(3) $x^2 + 2x + k^2 = 0$

(4) $kx^2 + 3x + 2 = 0$

(5) $2x^2 + 3x + k - 5 = 0$

(6) $x^2 + kx + (k - 1) = 0$

(7) $\frac{1}{3}x^2 + (k + 1)x + 8 = 0$

#11. Find the value of $p + q$ for the following quadratic equations with the solution $x = p \pm \sqrt{q}$

(1) $-2x^2 + 5x + 1 = 0$

(2) $3(x - 1)^2 = 4$

(3) $-(x + 1)^2 + 5 = 0$

#12. Find the constant a or the range of a for the following quadratic equations with a condition

(1) $(x + 1)^2 = a + 2$ has no solution.

(2) $x^2 + 3x + 3a = 0$ has two different solutions.

(3) $ax^2 + x + 2 = 0$ has one solution.

(4) $3x^2 - x + a = 0$ has no solution.

(5) $x^2 + (a + 1)x + \frac{a+3}{2} = 0$ has a double root.

#13. Find the value of the given expression for the following quadratic equations with two solutions (α, β)

(1) $\alpha\beta$ for $2(x + 3)^2 - 3 = 0$

(2) $\alpha + \beta$ for $-(x + 4)^2 + 5 = 0$

(3) $\alpha^2 + \beta^2$ for $3x^2 - x - 1 = 0$

#14. Find two constants (a and b) for the following quadratic equations with a condition

(1) $x^2 + x + a = 0$ has two different solutions, $x = 2$ and $x = b$.

(2) $x^2 + 2ax + b = 0$ has a double root $x = 3$.

(3) $x^2 + ax + b = 0$ has a solution $x = 1 + \sqrt{2}$.

(4) $x^2 - ax - 2b^2 = 0$ has two different solutions, $x = 4 \pm \sqrt{2a}$.

#15. Solve the following quadratic equations by using the quadratic formulas

(1) $x^2 - 2x - 4 = 0$

(2) $3x^2 + 5x - 1 = 0$

(3) $5x^2 - 2x - 1 = 0$

(4) $-2x^2 + 3x + 5 = 0$

(5) $\frac{1}{2}x^2 - 3x + 2 = 0$

(6) $\frac{1}{6}x^2 - 0.5x + \frac{1}{4} = 0$

(7) $(x + 1)^2 = 3(x + 2)$

(8) $(x + 2)^2 + 3(x + 2) - 2 = 0$

(9) $-\frac{(x-1)^2}{2} + x = 0.4(x + 1)$

(10) $(x + 3)(2x + 6) = 5$

#16. Find the value of the given expression for the following quadratic equations with a solution.

(1) $a + b$ for $(2x + 1)^2 = 3$ with $x = a \pm b\sqrt{3}$

(2) $a - b$ for $2x^2 - 8x + 1 = 0$ with $x = \frac{a \pm 3\sqrt{b}}{6}$

(3) ab for $ax^2 + 5x + 2 = 0$ with $x = \frac{-5 \pm 2\sqrt{b}}{4}$

(4) $\frac{b}{a}$ for $3x^2 - 5x + 1 = 0$ with $x = a \pm \sqrt{b}$

(5) $\frac{a+b}{ab}$ for $x^2 - 3x + 1 = 0$ with $x = \frac{a \pm 2\sqrt{b}}{2}$

(6) $\frac{a-b}{a^2-b^2}$ for $ax^2 + 3x - 3b = 0$ with $x = -1 \pm \sqrt{5}$

#17. Identify the number of solutions for each quadratic equation.

(1) $x^2 + 2x - 3 = 0$

(2) $-x^2 + x - 5 = 0$

(3) $4x^2 - 4x + 1 = 0$

(4) $kx^2 - (k + 5)x + 1 = 0$

(5) $3x^2 - x - k^2 = 0$

(6) $x^2 - 4kx + 5k^2 + 1 = 0$

#18. Find the value of a or range of a for the following quadratic equations with a condition

(Use the discriminant D)

(1) $x^2 + 5x + a = x + 2$ has no solution.

(2) $(a + 3)x^2 - 2ax + a - 1 = 0$ has two different solutions.

(3) $x^2 + 3ax - 2a + 3 = 0$ has only one solution.

(4) $x^2 + ax + a + 2 = 0$ has a double root and

$x^2 + 4ax + (2a - 1)^2 = 0$ has two different solutions.

(5) $2x^2 + (2a - 1)x + a^2 + \frac{1}{4} = 0$ has solutions.

(6) $(a - 1)x^2 + 2(a - 1)x + (a + 1) = 0$ has solutions.

#19. A quadratic equation $3x^2 + 5x - 2 = 0$ has two solutions, $x = \alpha$ and $x = \beta$.

Find the value of the given expressions.

(1) $\alpha + \beta$

(2) $\alpha^2 + \beta^2$

(3) $\alpha - \beta$

(4) $\alpha^2 - \beta^2$

(5) $\frac{1}{\alpha} + \frac{1}{\beta}$

#20. A quadratic equation $x^2 + 3kx + 2k^2 - 4k - 1 = 0$ has two solutions, $x = \alpha$ and $x = \beta$.

Find the value of the given expressions in terms of k for (1) through (5). Find the value of k for (6).

(1) $\alpha + \beta$

(2) $\alpha\beta$

(3) $\alpha^2 + \beta^2$

(4) $(\alpha - \beta)^2$

(5) $\frac{\beta}{\alpha} + \frac{\alpha}{\beta}$

(6) k if $\frac{1}{\alpha} + \frac{1}{\beta} = 1$

#21. Find the solution for the quadratic equation $ax^2 + (b - 1)x + 4 = 0$

(1) When the quadratic equation $2x^2 + (a - 1)x + b = 0$ has two solutions, $\frac{1}{2}$ and $\frac{1}{3}$.

(2) When the quadratic equation $3ax^2 + 8bx + 3 = 0$ has a double root -2.

(3) When the quadratic equation $ax^2 + 3ax - 4 = 0$ has two solutions, b and $b + 1$.

(4) When the quadratic equation $ax^2 + 3x + b = 0$ has two solutions, α and β, which satisfy the conditions $\alpha + \beta = -2$ and $\alpha\beta = 4$.

(5) When the quadratic equation $x^2 + ax + 3 = 0$ has two different solutions.

The one of the solutions is $x = -2 + 3\sqrt{b}$.

#22. $x^2 + ax + b = 0$ has two solutions, -2 and -3.

Find the value of $\alpha^2 + \beta^2$ for $x^2 - bx - a = 0$ which has solutions, α, β.

#23. Create a 1 x^2-coefficient quadratic equation which has two solutions, $\alpha + \beta$ and $\alpha\beta$, where α and β are both solutions of $x^2 + 2x - 3 = 0$.

#24. Find the range of a for the following quadratic equations with a condition

(1) The quadratic equation $x^2 - 3x + 2a = 0$ has two different positive solutions.

(2) The quadratic equation $ax^2 + 2x + 3 = 0$ has two different negative solutions.

(3) The quadratic equation $x^2 - 4x + 3a = 0$ has two different solutions (α and β) with opposite signs.

#25. The following quadratic equations have only one solution.

Find the solution (a double root) for each.

(1) $x^2 + kx + 2k - 3 = 0$

(2) $(k + 2)x^2 - 2kx + k + 1 = 0$

(3) $x^2 + (k + 2)x + k^2 - k + 2 = 0$

#26. An n-sided polygon has $\frac{n(n-3)}{2}$ diagonals. Find a polygon that has 20 diagonals.

#27. For three consecutive positive integers, the square of the biggest number is 12 less than the sum of the squares of the other numbers. Identify the biggest number.

#28. The product of two consecutive odd numbers is 99. Find the sum of the numbers.

#29. The sum of two positive numbers is 34 and their product is 225. Identify the two numbers.

#30. Nichole wants to produce a x^2% of salt solution after mixing 40 ounces of a 10% of salt solution with 40 ounces of a x% salt solution. Find the value of x.

#31. The difference between two positive integers is 2 and their product is 255.
Find the sum of the numbers.

#32. The area of a square A is 121 square inches.
Each side of square A is 3 inches longer than that of square B. Find the perimeter of the Square B.

#33. The perimeter and the area of a rectangle are 26 inches and 40 square inches, respectively.
Find the difference between the length and width of the rectangle's sides (in this case, length will be longer than width).

#34. Richard throws a ball upward with a beginning speed v of 60 feet per second.
The formula for the height in feet after t seconds is $h = vt - 5t^2$.

  (1) At what time will the height 100 feet?
  (2) When will the ball reach the ground again?
  (3) What height will the ball reach in 4 seconds?
  (4) What is the maximum height the ball will reach?

Chapter 14. Rational Expressions (Algebraic Functions)

14-1 Simplifying Rational Expressions

1. Rational Numbers and Rational Expressions

(1) *A rational number* is a fraction whose numerator and denominator are both integers and denoted by

$$\frac{a}{b} \text{ for any integers } a \text{ and } b(\neq 0).$$

(2) *A polynomial* is the sum of a number of terms, each of which is the product of numbers and letters. *Rational expression* is a fraction whose numerator and denominator are both polynomials and denoted by

$$\frac{P(x)}{Q(x)} \text{ for any polynomials } P(x) \text{ and } Q(x)(\neq 0).$$

Note   *Rational functions*

$y = \dfrac{1}{x}$ *is not defined at* $x = 0$.

$y = \dfrac{x+1}{x+2}$ *is not defined at* $x = -2$.

$y = \dfrac{x^2}{(x-1)(x+3)}$ *is not defined at either* $x = 1$ *or at* $x = -3$.

Note

$y = \dfrac{x}{x}$ *is not defined at* $x = 0$, *but for other variables of* x, $\dfrac{x}{x} = 1$.

Thus, the two expressions $\dfrac{x}{x}$ *and* 1 *are consequently not identical.*

Similarly, $y = \dfrac{x(x+1)}{x+1}$ *and* $y = x$ *are not the same rational functions.*

(∵ Both functions have the same variables for $x \neq -1$.

But $y = \dfrac{x(x+1)}{x+1}$ *is not defined at* $x = -1$, *whereas* $y = x$ *has the value* -1.)

2. Reducing Fractions to Lowest Terms

$$\frac{a^m}{a^n} = a^{m-n}, \quad a^{-m} = \frac{1}{a^m}$$

To reduce fractions to lowest terms, simplify the rational expressions and identify restrictions on the variables.

The domain of an expression is the set of real numbers for which the expression is defined.

Since division by zero is undefined, we must restrict the domain of the reduced expression by excluding the value, which would produce an undefined division by zero.

(1) When the quotients (fractions) of exponential expressions have the same base

To simplify the rational expression, reduce the fraction to its lowest term by the properties of the exponents. Then, reduce coefficients by prime factorizations.

Example

$$\frac{2x^3}{24x} = \frac{2 \cdot x^3}{2^3 \cdot 3 \cdot x} = \frac{x^2}{2^2 \cdot 3} = \frac{x^2}{12} \qquad \therefore \ \frac{2x^3}{24x} = \frac{x^2}{12}, \ x \neq 0$$

$$\frac{6x^2y^4}{15x^5y} = \frac{2 \cdot 3 \cdot x^2 \cdot y^4}{3 \cdot 5 \cdot x^5 y} = \frac{2}{5}x^{2-5}y^{4-1} = \frac{2}{5}x^{-3}y^3 = \frac{2y^3}{5x^3} \qquad \therefore \ \frac{6x^2y^4}{15x^5y} = \frac{2y^3}{5x^3}, \ y \neq 0$$

> $\frac{x^2}{x^2y} = \frac{1}{y} \ \Rightarrow$ Restriction $x \neq 0$ is required.
>
> However, $y \neq 0$ is not necessary
>
> ($\because$ If $y = 0$, then both of the expressions are undefined.)

(2) When the polynomials can be factorized

To simplify the rational expression, factor the numerator and the denominator. If the expression contains common factors, eliminate the common factors to reduce the fraction.

Example

(1) $\frac{2x^2-8}{x+2} = \frac{2(x^2-4)}{x+2} = \frac{2(x+2)(x-2)}{x+2} = 2(x-2)$

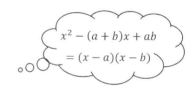

$x^2 - (a+b)x + ab$
$= (x-a)(x-b)$

    Therefore, $\frac{2x^2-8}{x+2} = 2(x-2), \ x \neq -2$

(2) $\frac{2x+4}{6x^2-18x-60} = \frac{2(x+2)}{6(x^2-3x-10)} = \frac{2(x+2)}{2 \cdot 3(x-5)(x+2)} = \frac{1}{3(x-5)}$

    Therefore, $\frac{2x+4}{6x^2-18x-60} = \frac{1}{3(x-5)}, \ x \neq -2$

(3) $\frac{2x^2-2}{6x^2+24x-30} = \frac{2(x^2-1)}{6(x^2+4x-5)} = \frac{2(x+1)(x-1)}{2 \cdot 3(x+5)(x-1)} = \frac{x+1}{3(x+5)}$

    Therefore, $\frac{2x^2-2}{6x^2+24x-30} = \frac{x+1}{3(x+5)}, \ x \neq 1$

3. Operations of Rational Expressions

(1) Adding and Subtracting Rational Expressions

For any polynomials $P(x), \ Q(x), \ R(x)(\neq 0)$, and $S(x)(\neq 0)$,

1) If the denominators are the same

Keep the denominator and add or subtract the numerators

$$\frac{P(x)}{R(x)} + \frac{Q(x)}{R(x)} = \frac{P(x)+Q(x)}{R(x)}, \quad \frac{P(x)}{R(x)} - \frac{Q(x)}{R(x)} = \frac{P(x)-Q(x)}{R(x)}$$

Example

$$\frac{3x}{x-2} + \frac{5}{x-2} = \frac{3x+5}{x-2} \quad , x \neq 2$$

$$\frac{x+1}{2x-1} - \frac{3x+5}{2x-1} = \frac{(x+1)-(3x+5)}{2x-1} = \frac{-2x-4}{2x-1} = \frac{-2(x+2)}{2x-1} \quad , x \neq \frac{1}{2}$$

2) If the denominators are different

Identify the least common denominator (LCD) of the fractions and simplify the expression by adding or subtracting the numerators, then dividing by the least common denominator.

$$\frac{P(x)}{R(x)} + \frac{Q(x)}{S(x)} = \frac{P(x)\cdot S(x)}{R(x)\cdot S(x)} + \frac{R(x)\cdot Q(x)}{R(x)\cdot S(x)} = \frac{P(x)\cdot S(x)+R(x)\cdot Q(x)}{R(x)\cdot S(x)},$$

$$\frac{P(x)}{R(x)} - \frac{Q(x)}{S(x)} = \frac{P(x)\cdot S(x)}{R(x)\cdot S(x)} - \frac{R(x)\cdot Q(x)}{R(x)\cdot S(x)} = \frac{P(x)\cdot S(x)-R(x)\cdot Q(x)}{R(x)\cdot S(x)}$$

Example

$$\frac{x-1}{2x(x+1)} + \frac{2}{3x(x-1)} = \frac{3\cdot(x-1)(x-1)}{6x(x+1)(x-1)} + \frac{2\cdot 2(x+1)}{6x(x-1)(x+1)}$$

> The least common denominator (LCD) is the least common multiple (LCM) of denominators.

$$= \frac{3(x^2-2x+1)+4(x+1)}{6x(x+1)(x-1)} = \frac{3x^2-2x+7}{6x(x+1)(x-1)}$$

$$\frac{3}{x+2} - \frac{2}{x-4} = \frac{3(x-4)}{(x+2)(x-4)} - \frac{2(x+2)}{(x-4)(x+2)} = \frac{3(x-4)-2(x+2)}{(x+2)(x-4)} = \frac{x-16}{(x+2)(x-4)}$$

(2) Multiplying and Dividing Rational Expressions

To multiply and divide rational expressions, common denominators are not necessary.

1) To multiply rational expressions,

Multiply the numerators, then multiply the denominators.

If possible, simplify the expressions before multiplying.

For any polynomials $P(x), Q(x), R(x)(\neq 0)$, and $S(x)(\neq 0)$,

$$\frac{P(x)}{R(x)} \cdot \frac{Q(x)}{S(x)} = \frac{P(x)\cdot Q(x)}{R(x)\cdot S(x)}$$

Example $\quad \dfrac{x+1}{3x} \cdot \dfrac{4x^2}{2(x+1)} = \dfrac{4(x+1)x^2}{3x\cdot 2(x+1)} = \dfrac{4\cdot(x+1)\cdot x\cdot x}{3\cdot x\cdot 2\cdot(x+1)} = \dfrac{2x}{3}$

$$\frac{2x^2-7x+3}{x^2+x-2} \cdot \frac{x^3+2x^2-3x}{4x^2-2x} = \frac{(2x-1)(x-3)}{(x+2)(x-1)} \cdot \frac{x(x+3)(x-1)}{2x(2x-1)} \quad \text{Factor and reduce}$$

$$= \frac{(x-3)(x+3)}{2(x+2)} \quad x \neq 0, \ x \neq 1, \ x \neq \frac{1}{2}$$

2) To divide rational expressions,

Convert the division into multiplication using the reciprocal of the 2nd fraction (the fraction we are dividing by). If possible, simplify the expressions before multiplying.

For any polynomials $P(x), Q(x), R(x)(\neq 0)$, and $S(x)(\neq 0)$,

$$\frac{P(x)}{R(x)} \div \frac{Q(x)}{S(x)} = \frac{P(x)}{R(x)} \cdot \frac{S(x)}{Q(x)} = \frac{P(x) \cdot S(x)}{R(x) \cdot Q(x)}, Q(x)(\neq 0)$$

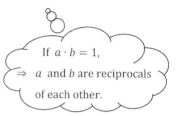

If $a \cdot b = 1$,
$\Rightarrow$ a and b are reciprocals of each other.

Example

$$\frac{2x}{(x+1)^2} \div \frac{1}{x^2-2x-3} = \frac{2x}{(x+1)^2} \cdot \frac{x^2-2x-3}{1} = \frac{2x(x^2-2x-3)}{(x+1)^2} = \frac{2x(x-3)(x+1)}{(x+1)(x+1)} = \frac{2x(x-3)}{(x+1)}$$

3) To divide long rational expressions (Long division of polynomials),

Let $P(x)$ and $D(x)$ be polynomials of degrees n and r, respectively, where $n \geq r$.

There exist polynomials $Q(x)$, called the *quotient*, and $R(x)$, called the *remainder*, such that:

① $P(x) = D(x) \cdot Q(x) + R(x)$ for all x

② The degree of the remainder $R(x)$ must be less than the degree of the divisor $D(x)$,

that is, $\deg R(x) < \deg D(x)$ or $R(x) = 0$

Note : If the degree of the divisor $D(x)$ is 1, then the degree of the remainder $R(x)$ is 0,

i.e., $R(x) = a$ (a is a constant).

If the degree of the divisor $D(x)$ is 2, then the degree of the remainder $R(x)$ is 1,

i.e., $R(x) = ax + b$ ($a \neq 0$).

If the degree of the divisor $D(x)$ is 3, then the degree of the remainder $R(x)$ is 2,

i.e., $R(x) = ax^2 + bx + c$ ($a \neq 0$).

Example

Divide $x^3 + 2x^2 + 4x + 5$ by $x + 2$ using polynomial long division.

$$\frac{x^3+2x^2+4x+5}{x+2}$$

$$\Rightarrow \quad \begin{array}{r} x^2 \qquad + 4 \\ x+2 \overline{) x^3 + 2x^2 + 4x + 5} \\ \underline{x^3 + 2x^2} \\ 4x + 5 \\ \underline{4x + 8} \\ -3 \end{array}$$

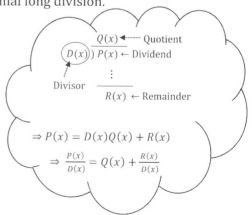

$Q(x) \leftarrow$ Quotient
$D(x) \overline{) P(x)} \leftarrow$ Dividend
Divisor $\quad \vdots$
$R(x) \leftarrow$ Remainder

$\Rightarrow P(x) = D(x)Q(x) + R(x)$

$\Rightarrow \frac{P(x)}{D(x)} = Q(x) + \frac{R(x)}{D(x)}$

$$\therefore \frac{x^3+2x^2+4x+5}{x+2} = x^2 + 4 - \frac{3}{x+2}$$

Note   If terms are missing from the dividend or the divisor, supply the missing terms with zero coefficients.

Example

$\dfrac{4x^3+6x^2-3}{2x-1}$

$\Rightarrow \dfrac{4x^3+6x^2+0\cdot x-3}{2x-1}$

$$\Rightarrow \begin{array}{r} 2x^2 + 4x\ + 2 \\ 2x-1\ )\overline{\ 4x^3 + 6x^2 + 0\cdot x - 3\ } \\ \underline{4x^3 - 2x^2\qquad\qquad} \\ 8x^2 + 0\cdot x - 3 \\ \underline{8x^2 - 4x\qquad} \\ 4x\ - 3 \\ \underline{4x\ - 2} \\ -1 \end{array}$$

$\therefore\ \dfrac{4x^3+6x^2-3}{2x-1} = 2x^2 + 4x + 2 - \dfrac{1}{2x-1}$

4) To simplify complex fractions,

Convert the complex fraction into a rational quotient. Rewrite the rational quotient as a product and simplify the expressions.

$\dfrac{\frac{a}{b}}{\frac{c}{d}} = \dfrac{ad}{bc}$

Example

$\dfrac{\frac{x(x+1)}{x^2}}{\frac{x^2+6x+5}{x-2}} = \dfrac{x(x+1)}{x^2} \div \dfrac{x^2+6x+5}{x-2} = \dfrac{x(x+1)}{x^2} \cdot \dfrac{x-2}{x^2+6x+5} = \dfrac{x\cdot(x+1)\cdot(x-2)}{x\cdot x\cdot(x+5)\cdot(x+1)}$

$\qquad = \dfrac{x-2}{x(x+5)}\ ,\ x \neq 0,\ x \neq -1$

5) Synthetic Division

When the divisor is a linear binomial (of the form $(x + \text{constant})$), the long division of polynomials is solved by synthetic division.

Example   Divide $x^3 + 2x^2 + 4x + 5$ by $x + 2$ using synthetic division.

Write the coefficients (1, 2, 4, 5) of the dividend, $x^3 + 2x^2 + 4x + 5$, in a line

$$\begin{array}{r|ccc} & 1 & 2 & 4 & 5 \\ \hline & & & & \end{array}$$

To the left side of the coefficients, write the opposite (-2) of the constant term of the divisor, $x + 2$

$$-2 \begin{array}{|cccc} 1 & 2 & 4 & 5 \end{array}$$

Bring first coefficient (1) below the line

$$-2 \begin{array}{|cccc} 1 & 2 & 4 & 5 \\ \hline 1 \end{array}$$

Multiply the left number (-2) by the number (1) below the line and place the product $(-2 \cdot 1 = -2)$ under the second coefficient (2)

$$-2 \begin{array}{|cccc} 1 & 2 & 4 & 5 \\ & -2 \\ \hline 1 \end{array}$$

Add the second coefficient (2) with the product (-2). Write the sum below the line to complete the column

$$-2 \begin{array}{|cccc} 1 & 2 & 4 & 5 \\ & -2 \\ \hline 1 & 0 \end{array}$$

Multiply the left number (-2) by the number (0) and place the product $(-2 \cdot 0 = 0)$ under the third coefficient (4). Add the third coefficient (4) and the product (0). Write the sum below the line to complete the column:

$$-2 \begin{array}{|cccc} 1 & 2 & 4 & 5 \\ & -2 & 0 \\ \hline 1 & 0 & 4 \end{array}$$

Repeat the steps, multiply, place, and add.

$$-2 \begin{array}{|cccc} 1 & 2 & 4 & 5 \\ & -2 & 0 & -8 \\ \hline 1 & 0 & 4 & -3 \end{array}$$

Coefficients of quotient

All the numbers below the line except the last number (-3) represent the coefficients of the quotient. The last number (-3) represents the remainder. Since we divided the dividend by a linear binomial which has a power of 1, the power of the quotient is 1 less than the power of the dividend. Since the dividend has a power of 3, the quotient must have a power of 2. Therefore, the quotient is $1 \cdot x^2 + 0 \cdot x + 4$ and remainder is -3.

That is, $\dfrac{x^3+2x^2+4x+5}{x+2} = x^2 + 4 - \dfrac{3}{x+2}$

$x + 2 = x^{①}+ 2$; power of 1

$x^{③} + 2x^2 + 4x + 5$; power of 3

Example Divide $4x^3 + 6x^2 - 3$ by $2x - 1$ using synthetic division.

Since the divisor is $2x - 1$, express it as the form of $x +$constant (linear binomial).

Dividing the expression by 2, we get $\dfrac{4x^3+6x^2-3}{2x-1} = \dfrac{2x^3+3x^2-\frac{3}{2}}{x-\frac{1}{2}}$.

Now we can use synthetic division.

Since x-term is missing, supply the x-term with a zero coefficient.

$$
\begin{array}{c|cccc}
\frac{1}{2} & 2 & 3 & 0 & -\frac{3}{2} \\
 & & 1 & 2 & 1 \\
\hline
 & 2 & 4 & 2 & -\frac{1}{2}
\end{array}
$$

$\therefore \dfrac{2x^3+3x^2-\frac{3}{2}}{x-\frac{1}{2}} = 2x^2 + 4x + 2 - \dfrac{\frac{1}{2}}{x-\frac{1}{2}} = 2x^2 + 4x + 2 - \dfrac{1}{2x-1}$

Therefore, $\dfrac{4x^3+6x^2-3}{2x-1} = 2x^2 + 4x + 2 - \dfrac{1}{2x-1}$

14-2 Solving Rational Equations and Inequalities

1. Rational Equations

(1) Using the Cross Product (Multiplication)

In a proportion,

$$\dfrac{P(x)}{R(x)} = \dfrac{Q(x)}{S(x)} \implies P(x)S(x) = Q(x)R(x), \text{ where } R(x) \neq 0, S(x) \neq 0$$

A proportion is an equality of two ratios.

Example

$\dfrac{20}{x^2+4x+3} = \dfrac{x}{x+3}$

$\implies \dfrac{20}{(x+1)(x+3)} = \dfrac{x}{x+3}$

$\implies x(x + 1)(x + 3) = 20(x + 3) \implies x(x + 1) = 20$

$\implies x^2 + x - 20 = 0 \implies (x + 5)(x - 4) = 0 \implies x = -5 \text{ or } x = 4$

$\therefore$ The solution to the equation is $x = -5$ or $x = 4$

(2) Using the Least Common Denominator (LCD)

If a rational equation consists of two or more fractions on one side of the equal sign, combine the expressions on the side containing fractions by using their least common denominator (LCD). Then simplify the expression.

Example

$3 + \dfrac{3}{x-2} = -2x$

$\Rightarrow \dfrac{3(x-2)}{x-2} + \dfrac{3}{x-2} = -2x$ The LCD of $\dfrac{3}{1}$ and $\dfrac{3}{x-2}$ is $x-2$

$\Rightarrow \dfrac{3x-3}{x-2} = -2x$ Combine the expressions on the left side of the equation

$\Rightarrow -2x(x-2) = 3x-3$ Cross multiply to eliminate a fraction

$\Rightarrow 2x^2 - x - 3 = 0$

$\Rightarrow (2x-3)(x+1) = 0$ Factor the quadratic equation

$\Rightarrow 2x - 3 = 0 \text{ or } x + 1 = 0$

$\Rightarrow x = \dfrac{3}{2} \text{ or } x = -1$

Therefore, the solution to the equation is $x = \dfrac{3}{2}$ or $x = -1$.

2. Rational Inequalities

(1) Critical Numbers

If the denominator of a rational expression is equal to zero, the expression is not defined. Any values by which the expression is undefined or equal to zero are called the *critical numbers* of the rational expression. These critical numbers divide a number line into intervals.

(2) Solving Rational Inequalities

To solve rational inequalities, choose a test value from each interval to identify the solutions.

Example

$\dfrac{x-1}{x+2} > 0$

If the denominator equals zero ($x + 2 = 0$), then $x = -2$, and the expression is undefined.

If the numerator equals zero ($x - 1 = 0$), then $x = 1$, and the expression is equal to zero.

Thus, the critical numbers of the rational expression are $x = -2$ and $x = 1$.

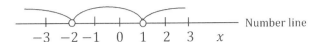

Number line

These critical numbers divide the number line into three intervals.

Case 1. $x < -2$

> $\Rightarrow$ Choose a test value, such as $x = -3$, and substitute the test value into the
>
> expression. Then, $\dfrac{x-1}{x+2} = \dfrac{-3-1}{-3+2} = \dfrac{-4}{-1} = 4$
>
> Since $4 > 0$, $\dfrac{x-1}{x+2} > 0$ is true.

Case 2. $-2 < x < 1$

> $\Rightarrow$ Choose a test value, such as $x = 0$, and substitute the test value into the
>
> expression. Then, $\dfrac{x-1}{x+2} = \dfrac{0-1}{0+2} = -\dfrac{1}{2}$
>
> Since $-\dfrac{1}{2} < 0$, $\dfrac{x-1}{x+2} > 0$ is false.

Case 3. $x > 1$

> $\Rightarrow$ Choose a test value, such as $x = 2$, and substitute the test value into the
>
> expression. Then, $\dfrac{x-1}{x+2} = \dfrac{2-1}{2+2} = \dfrac{1}{4}$
>
> Since $\dfrac{1}{4} > 0$, $\dfrac{x-1}{x+2} > 0$ is true.

Therefore, the solution to the rational inequality is $x < -2$ or $x > 1$.

Note   *If $\dfrac{P(x)}{Q(x)} > 0$ or $\dfrac{P(x)}{Q(x)} < 0$, then the critical numbers are open points on the number line.*

Note   *If $\dfrac{P(x)}{Q(x)} \geq 0$ or $\dfrac{P(x)}{Q(x)} \leq 0$, then the critical numbers are closed points on the number line.*

> *However, the number for which the denominator equals zero (which makes the expression*
>
> *undefined) is always an open point on a number line.*

When $a > b$,

① If $(x - a)(x - b) > 0$ (positive)

$\Rightarrow x > a$ or $x < b$ x is greater than the larger number or
                                  x is less than the smaller number

② If $(x - a)(x - b) < 0$ (negative)

$\Rightarrow b < x < a$ x is between the smaller number
                      and the larger number

$\therefore \dfrac{x-1}{x+2} > 0 \Rightarrow (x - 1)(x + 2) > 0$

$\Rightarrow x > 1$ or $x < -2$

Example

$$\frac{x}{3} \leq \frac{5}{x+2}$$

$$\Rightarrow \frac{x}{3} - \frac{5}{x+2} \leq 0$$

$$\Rightarrow \frac{x(x+2)-15}{3(x+2)} \leq 0$$

$$\Rightarrow \frac{x^2+2x-15}{3(x+2)} \leq 0$$

$$\Rightarrow \frac{(x+5)(x-3)}{3(x+2)} \leq 0$$

The critical numbers are $x = -2$, $x = -5$, and $x = 3$.

The number line is now divided into four intervals.

If $x = -2$, then the expression is undefined.

So, $x = -2$ is not included. We draw an open point for this value on the number line.

Case1. $x \leq -5 \Rightarrow$ try $x = -6$: $\frac{-6}{3} \leq \frac{5}{-6+2}$; $-2 \leq \frac{5}{-4}$ (True)

Case2. $-5 \leq x < -2 \Rightarrow$ try $x = -3$: $\frac{-3}{3} \leq \frac{5}{-3+2}$; $-1 \leq -5$ (False)

Case3. $-2 < x \leq 3 \Rightarrow$ try $x = 0$: $\frac{0}{3} \leq \frac{5}{0+2}$; $0 \leq \frac{5}{2}$ (True)

Case4. $x \geq 3 \Rightarrow$ try $x = 6$: $\frac{6}{3} \leq \frac{5}{6+2}$; $2 \leq \frac{5}{8}$ (False)

Therefore, the solution to the rational inequality is $x \leq -5$ or $-2 < x \leq 3$.

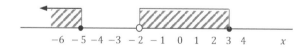

Exercises

#1. Simplify each expression.

(1) $\dfrac{2x^5}{x^2}$

(2) $\dfrac{12x^4y}{56x^2y^6}$

(3) $\dfrac{8-2x}{3x-12}$

(4) $\dfrac{x^2+6x+8}{x^2+3x-4}$

(5) $\dfrac{3}{x}+\dfrac{1}{2x}$

(6) $\dfrac{2x}{x^2-4}-\dfrac{3}{x+2}$

(7) $\dfrac{x+1}{x^2-9}-\dfrac{x}{x^2-2x-15}$

(8) $\dfrac{x+3}{x^2-4}\cdot\dfrac{x+2}{x^2+2x-3}$

(9) $\dfrac{x^2-16}{(x+5)^2}\div\dfrac{x+4}{x^2+2x-15}$

(10) $\dfrac{\frac{2}{x^3}}{\frac{8}{x^5}}$

(11) $\dfrac{\frac{2x^2-5x-3}{x-3}}{\frac{2x^2-7x-4}{2x}}$

#2. Divide the following long rational expressions

(1) $(x^2+5x-6)\div(x-2)$

(2) $(2x^2-20)\div(x+3)$

(3) $(3x^3+2x^2+1)\div(x-2)$

#3. Solve the following rational equations

(1) $\dfrac{3x}{2}=\dfrac{5}{8}$

(2) $\dfrac{2}{3x-4}=\dfrac{-1}{2-4x}$

(3) $7+\dfrac{4}{x-1}=-2x$

(4) $\dfrac{1}{x+2}+\dfrac{2}{x^2-4}=\dfrac{4}{x-2}$

#4. Solve the following inequalities

(1) $\dfrac{x-4}{x+3}>0$

(2) $\dfrac{x+5}{x-2}\le 0$

(3) $\dfrac{x}{2}\ge\dfrac{5}{x-3}$

(4) $\dfrac{2x^2-5x-3}{x-4}>0$

Chapter 15. Quadratic Functions

15-1 Quadratic Functions and their Graphs

1. Quadratic Functions

> $ax^2 + bx + c = 0,\ a \neq 0$: quadratic equation
>
> $y = ax^2 + bx + c,\ a \neq 0$: quadratic function

(1) Definition

For any function $y = f(x)$, a function represented by the form

$$y = ax^2 + bx + c, \text{ with constants } a \neq 0,\ b,\ c$$

is called a *quadratic function*.

> Quadratic :
> The highest power in a polynomial is 2.

Example $y = 3x^2,\ y = -x^2 + 1,\ y = 2x^2 + 3x,\ y = \frac{1}{2}x^2 - x + 2$

(2) Parabola

1) A graph of a quadratic function $y = ax^2 + bx + c, a \neq 0$ on a coordinate plane
   is called a *parabola*.

2) The symmetrical axis (central line) of a parabola is called the *axis* of the parabola or *the axis of symmetry*.

   A parabola is symmetrical with respect to the axis of symmetry.

3) The intersection point of the parabola and the axis of symmetry is called the *vertex*.

4) There are two types of parabola:

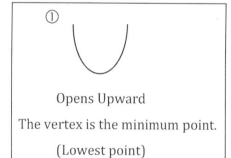

①

Opens Upward

The vertex is the minimum point.

(Lowest point)

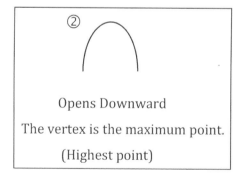

②

Opens Downward

The vertex is the maximum point.

(Highest point)

Note: On the graph of the simplest quadratic equation $y = x^2$, the equation of the axis of symmetry is the line $x = 0$ (y axis), and the coordinates of the vertex are $(0,0)$.

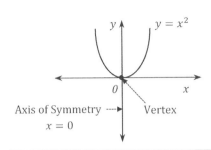

Axis of Symmetry $\cdots\blacktriangleright$ Vertex

$x = 0$

The graph of a linear function = Line
The graph of a quadratic function= Parabola

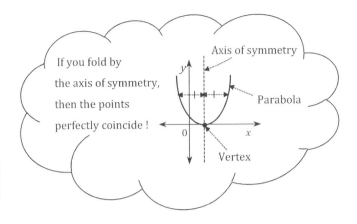

If you fold by
the axis of symmetry,
then the points
perfectly coincide !

2. Graphing of Quadratic Functions

(1) Graphing $y = ax^2$, $a \neq 0$

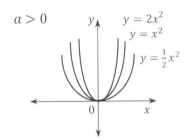

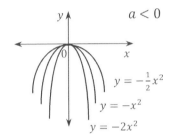

① Vertex $(0,0)$

② Axis of symmetry $x = 0$ (y-axis)

③ $a > 0 \Rightarrow$ Open upward and $y = f(x) \geq 0$

     $a < 0 \Rightarrow$ Open downward and $y = f(x) \leq 0$

> The point (a, b) is on the graph of $y = f(x)$.
> $\Rightarrow f(a) = b$

④ The larger the magnitude of the absolute value of a, the narrower the width of the

     parabola.

Note The graphs of $y = ax^2$ and $y = -ax^2$ are symmetric along the x-axis.

<u>When $a > 0$,</u>
$x \uparrow \Rightarrow f(x) \uparrow$
(If the value of x increases, then
the value of $f(x)$ increases.)

<u>When $a < 0$,</u>
$x \uparrow \Rightarrow f(x) \downarrow$
(If the value of x increases, then
the value of $f(x)$ decreases.)

(2) Graphing $y = ax^2 + q$, $a \neq 0$

1) $a > 0$, $q > 0$

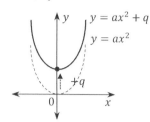

2) $a < 0$, $q > 0$

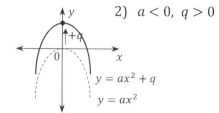

The graph of $y = ax^2 + q$ is a translation of the graph of $y = ax^2$,

with the q units along the y-axis.

① Vertex $(0, q)$

② Axis of symmetry $x = 0$ (y-axis)

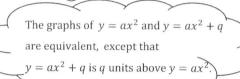

The graphs of $y = ax^2$ and $y = ax^2 + q$ are equivalent, except that $y = ax^2 + q$ is q units above $y = ax^2$.

(3) Graphing $y = a(x - p)^2$, $a \neq 0$

1) $a > 0$, $p > 0$

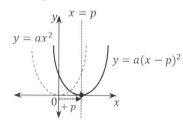

2) $a < 0$, $p > 0$

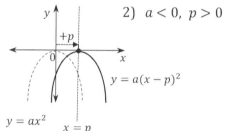

The graph of $y = a(x - p)^2$ is a translation of the graph of $y = ax^2$,

with the p units along the x-axis.

① Vertex $(p, 0)$

② Axis of symmetry $x = p$

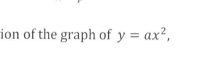

Shift the graph of $y = ax^2$ p units to the right and q units up.

(4) Graphs of $y = a(x - p)^2 + q$, $a \neq 0$

1) $a > 0$, $p > 0$, $q > 0$

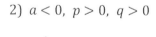

2) $a < 0$, $p > 0$, $q > 0$

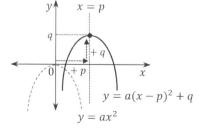

The graph of $y = a(x - p)^2 + q$ is a translation of the graph of $y = ax^2$,

with the p units along the x-axis and the q units along the y-axis.

① Vertex (p, q)

② Axis of symmetry $x = p$

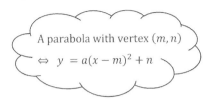

A parabola with vertex (m, n)

$\Leftrightarrow y = a(x - m)^2 + n$

Note : Graph of $y = a(x + p)^2 - q$, $a \neq 0$

$a < 0$, $p > 0$, $q > 0$

The graph of $y = a(x + p)^2 - q$ *is a translation*

of the graph of $y = ax^2$, *with the* $-p$ *units along the* x-axis

and the $-q$ *units along the* y-axis.

① *Vertex* $(-p, -q)$

② *Axis of symmetry* $x = -p$

15-2 Properties of Quadratic Functions

1. Graphing $y = ax^2 + bx + c$

To find the vertex and the axis of symmetry for a graph, transform the graph of the general

quadratic function form $y = ax^2 + bx + c$, $a \neq 0$ to the graph of a standard form:

$$y = a(x - p)^2 + q, \ a \neq 0$$

Note: $\quad y = ax^2 + bx + c = a\left(x^2 + \dfrac{b}{a}x\right) + c = a\left(\left(x + \dfrac{b}{2a}\right)^2 - \left(\dfrac{b}{2a}\right)^2\right) + c$

$$= a\left(x + \dfrac{b}{2a}\right)^2 - \dfrac{b^2}{4a} + c = a\left(x + \dfrac{b}{2a}\right)^2 - \dfrac{b^2 - 4ac}{4a}$$

(1) $y = ax^2 + bx + c \xrightarrow[\text{transform}]{} y = a\left(x + \dfrac{b}{2a}\right)^2 - \dfrac{b^2 - 4ac}{4a}$

(2) Vertex $\left(-\dfrac{b}{2a}, -\dfrac{b^2 - 4ac}{4a}\right)$

(3) Axis of symmetry $x = -\dfrac{b}{2a}$

2. Intercepts

(1) x-intercept

The x-intercept is the x-coordinate of the point where the graph intersects the x-axis.

x-intercept is the value of x when $y = 0$.

(2) *y*-intercept

The *y-intercept* is the *y*-coordinate of the point where the graph intersects the *y*-axis.

y-intercept is the value of *y* when $x = 0$.

Note:

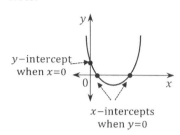

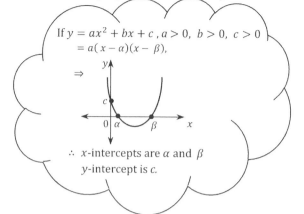

If $y = ax^2 + bx + c$, $a > 0$, $b > 0$, $c > 0$
$= a(x - \alpha)(x - \beta)$,

$\Rightarrow$

∴ *x*-intercepts are α and β
y-intercept is *c*.

Note

(1) *If* $y = ax^2 + bx + c = a(x - \alpha)(x - \beta)$,

  the equation $ax^2 + bx + c = 0$ *has two different solutions.*

  (*using* $D = b^2 - 4ac > 0$)

  $\Rightarrow$ *The parabola has* 2 *different x-intercepts.*

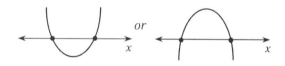

or

(2) *If* $y = ax^2 + bx + c = a(x - \alpha)^2$,

  the equation $ax^2 + bx + c = 0$ *has one solution (the double root).*

  (*using* $D = b^2 - 4ac = 0$)

  $\Rightarrow$ *The parabola has only* 1 *x-intercept.*

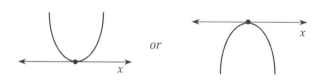

or

(3) *If* $y = ax^2 + bx + c$,

  the equation $ax^2 + bx + c = 0$ *has no solution.*

  (*using* $D = b^2 - 4ac < 0$)

  $\Rightarrow$ *The parabola has no x-intercept.*

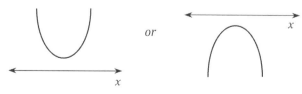

or

Example Find an equation for the parabola that has its vertex at $(2, 4)$ and that passes

through the origin $(0, 0)$.

The standard form of the parabola with vertex $(2, 4)$ is $f(x) = a(x - 2)^2 + 4$

Since the parabola passes through the point $(0, 0)$, $f(0) = 0$.

$\therefore\ 0 = a(0 - 2)^2 + 4 = 4a + 4 \quad \therefore\ a = -1$

Thus, the equation is $f(x) = -(x - 2)^2 + 4 = -x^2 + 4x$

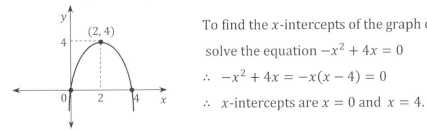

To find the x-intercepts of the graph of $f(x) = -x^2 + 4x$,

solve the equation $-x^2 + 4x = 0$

$\therefore\ -x^2 + 4x = -x(x - 4) = 0$

$\therefore\ x$-intercepts are $x = 0$ and $x = 4$.

3. Translation and Symmetry

(1) $y = ax^2$

1) Translation

To find the translation of the graph of $y = ax^2$,

with the p units along the x-axis and the q units along the y-axis.

$\Rightarrow$ Substitute $x - p$ into x and substitute $y - q$ into y.

$\Rightarrow\ y - q = a(x - p)^2$

$\Rightarrow\ y = a(x - p)^2 + q$

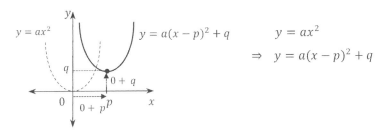

$$y = ax^2$$
$$\Rightarrow\ y = a(x - p)^2 + q$$

2) Symmetry

① To create symmetry along the x-axis: substitute $-y$ into y.

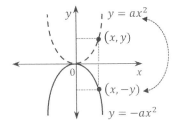

$$y = ax^2$$
$$\Rightarrow\ -y = ax^2$$
$$\Rightarrow\ y = -ax^2$$

② To create symmetry along the y-axis: substitute $-x$ into x.

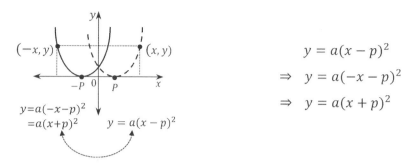

$$y = a(x - p)^2$$
$$\Rightarrow \ y = a(-x - p)^2$$
$$\Rightarrow \ y = a(x + p)^2$$

(2) $y = a(x - p)^2 + q$

1) Translation

To find the translation of the graph of $y = a(x - p)^2 + q$,

with the m units along the x-axis and the n units along the y-axis :

$\Rightarrow$ Substitute $x - m$ into x and $y - n$ into y.

$\Rightarrow y - n = a(x - m - p)^2 + q$

$\Rightarrow y = a(x - (p + m))^2 + q + n$

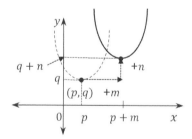

$$y = a(x - p)^2 + q$$
$$\Rightarrow y = a(x - (p + m))^2 + (q + n)$$

2) Symmetry

① To create symmetry along the x-axis: Substitute $-y$ into y.

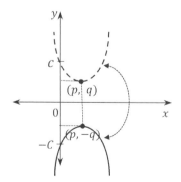

$$y = a(x - p)^2 + q$$
$$\Rightarrow \ -y = a(x - p)^2 + q$$
$$\Rightarrow \quad y = -a(x - p)^2 - q$$

② To create symmetry along the y-axis: Substitute $-x$ into x.

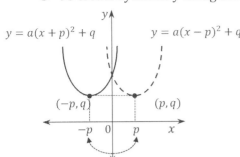

$$y = a(x - p)^2 + q$$
$$\Rightarrow y = a(-x - p)^2 + q$$
$$\Rightarrow y = a(x + p)^2 + q$$

4. Conditions for the y-value

$y = ax^2 + bx + c = a(x - p)^2 + q, \ a \neq 0$ Standard form

(1) $a > 0$

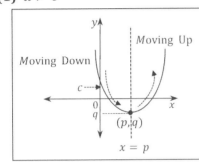

i) When $x < p$,

   If the $x-$values are increasing , $\Rightarrow$ the $y-$values are decreasing.
     (Moving right)   (Moving down)

ii) When $x > p$,

   If the $x-$values are increasing, $\Rightarrow$ the $y-$values are increasing.
     (Moving right)   (Moving up)

(2) $a < 0$

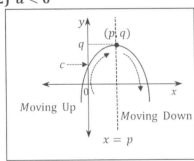

i) When $x < p$,

   If the $x-$values are increasing, $\Rightarrow$ the $y-$values are increasing.
     (Moving right)   (Moving up)

ii) When $x > p$,

   If the $x-$values are increasing, $\Rightarrow$ the $y-$values are decreasing.
     (Moving right)   (Moving down)

5. Properties of $y = a(x - p)^2 + q$

(1) The signs of the x^2-coefficient

1) $a > 0$

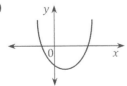

$\Rightarrow$ Opens upward

2) $a < 0$

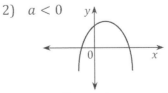

$\Rightarrow$ Opens downward

$y = ax^2 + bx + c$

① Sign of a:

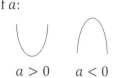

$a > 0 \qquad a < 0$

② Sign of b:

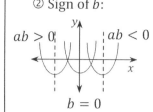

$ab > 0 \qquad ab < 0$

$b = 0$

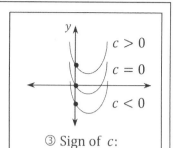

$c > 0$

$c = 0$

$c < 0$

③ Sign of c:

(2) The sides of the axis of Symmetry

$y = a(x - p)^2 + q$, $a > 0$ $\Rightarrow$ Axis of symmetry $x = p$

① $p > 0$

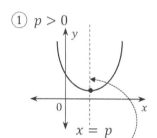

$x = p$

On the right side of the y-axis

② $p < 0$

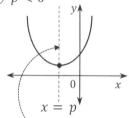

$x = p$

On the left side of the y-axis

③ $p = 0$

$\Rightarrow$ The y-axis is the axis of symmetry.

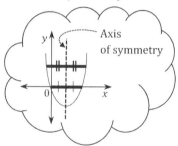

Axis of symmetry

Note: $y = ax^2 + bx + c$ $\Rightarrow$ $y = a\left(x + \dfrac{b}{2a}\right)^2 - \dfrac{b^2 - 4ac}{4a}$

$\Rightarrow$ Axis of symmetry is $x = -\dfrac{b}{2a}$

i) If $-\dfrac{b}{2a} > 0$ ($\dfrac{b}{2a} < 0$; $\dfrac{b}{a} < 0$; a and b have different signs),

then the axis of symmetry is on the right side of the y-axis

ii) If $-\dfrac{b}{2a} < 0$ ($\dfrac{b}{2a} > 0$; $\dfrac{b}{a} > 0$; a and b have the same sign),

then the axis of symmetry is on the left side of the y-axis

iii) If $-\dfrac{b}{2a} = 0$, then $b = 0$

(3) The signs of the y-intercept

If $y = ax^2 + bx + c = a(x - p)^2 + q$, $a > 0$, then the y-intercept is c. $(c = ap^2 + q)$

① $c > 0$	② $c < 0$	③ $c = 0$
$\Rightarrow$ The $y-$intercept is positive.	$\Rightarrow$ The $y-$intercept is negative.	$\Rightarrow$ The $y-$intercept is 0.
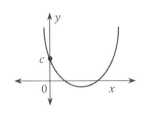	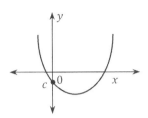	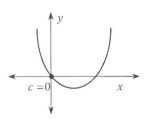
Above the x-axis	Below the x-axis	Pass through the origin

15-3 Solving Quadratic Functions

1. Equations of Quadratic Functions

(1) If the vertex (p, q) and one point are given,

$$\Rightarrow \quad y = a(x - p)^2 + q$$

Find a by substituting the point in the equation.

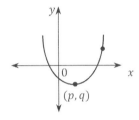

(2) If the axis of symmetry $x = p$ and two different points are given,

$$\Rightarrow \quad y = a(x - p)^2 + q$$

Find a and q

by substituting the two points in the equation.

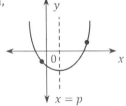

(3) If three different points are given,

$$\Rightarrow \quad y = ax^2 + bx + c$$

Find a, b and c

by substituting the points in the equation.

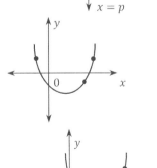

(4) If the x-intercepts α, β and one point are given,

$$\Rightarrow \quad y = a(x - \alpha)(x - \beta)$$

Find a by substituting the point in the equation.

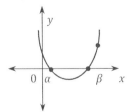

2. Maximum or Minimum Values of Quadratic Functions

To get the maximum value or the minimum value of a quadratic function, transform the general form of the quadratic function $y = ax^2 + bx + c$ to the standard form of a quadratic function $y = a(x - p)^2 + q$.

(1) $y = a(x - p)^2 + q$

1) $a > 0$

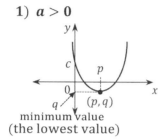

minimum value
(the lowest value)

When $x = p$,

$y = q$ is the minimum value of this quadratic function. All the y-values of this function are greater than or equal to q ($f(x) \geq q$). Since the graph opens upward, there is no maximum value.

> If $a > 0 \Rightarrow$ the minimum at the vertex, there is no maximum
>
> If $a < 0 \Rightarrow$ the maximum at the vertex, there is no minimum

2) $a < 0$

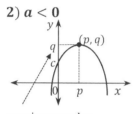

maximum value
(the highest value)

When $x = p$,

$y = q$ is the maximum value of this quadratic function. All the y-values of this function are less than or equal to q ($f(x) \leq q$). Since the graph opens downward, there is no minimum value.

Note: $y = a(x - \alpha)(x - \beta)$

$\Rightarrow$ *This parabola has its maximum or minimum value at* $x = \dfrac{\alpha + \beta}{2}$

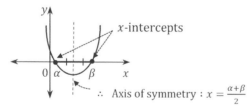

x-intercepts

$\therefore$ Axis of symmetry : $x = \dfrac{\alpha + \beta}{2}$

(2) $y = ax^2 + bx + c$

> First, transform to the standard form !

$$y = ax^2 + bx + c = a\left(x + \frac{b}{2a}\right)^2 - \frac{b^2 - 4ac}{4a}$$

1) $a > 0 \Rightarrow$ minimum value: $-\dfrac{b^2 - 4ac}{4a}$, when $x = -\dfrac{b}{2a}$

2) $a < 0 \Rightarrow$ maximum value: $-\dfrac{b^2 - 4ac}{4a}$, when $x = -\dfrac{b}{2a}$

(3) When a Maximum Value or Minimum Value is given:

    1) The minimum value m at $x = n$

$$\Rightarrow y = a(x - n)^2 + m, \qquad a > 0$$

    2) The maximum value m at $x = n$

$$\Rightarrow y = a(x - n)^2 + m, \qquad a < 0$$

3. Steps for Solving Word Problems

(1) Determine the two variables x and y.

(2) Express the relationship between the variables x and y by considering the range of x-value.

(3) Simplify the function.

(4) Using the graph of the function, find the solution.

(5) Check the solution.

Exercises

#1 Identify the quadratic functions by marking O or $\times$.

(1) $y = \frac{1}{2}x^2 + 1$

(2) $y = 2x^2 - (3 + 2x^2)$

(3) $y = \frac{1}{x^2} + 1$

(4) $y = x^2 - (x + 1)^2$

(5) $y = x(x + 1)$

(6) $y = 2x^2 - x^2 + 1$

(7) $y = \frac{(x+1)^2}{3}$

(8) $2x^2 + 3x + 1$

(9) $y = 2$

(10) $y = 3x + 1$

#2 Find the following values of the quadratic function $f(x) = x^2 - 2x - 1$.

(1) $f(0)$

(2) $f(-1)$

(3) $f(2) + f(-2)$

(4) $f\left(-\frac{1}{2}\right)$

(5) $2f(1)$

#3 Find the value of $a + b$ for the quadratic equation $f(x) = -\frac{1}{2}x^2 + a$.

(1) $f(1) = -3$ and $f(-2) = b$

(2) $f(-1) = 1$ and $\frac{1}{2}f(0) = 2b$

(3) $\frac{f(1)+f(-1)}{2} = -\frac{1}{4}$ and $f(2) = -b$

#4 Find the vertex and the axis of symmetry for the following parabolas

(1) $y = 2x^2 - 4x$

(2) $y = x^2 - 2x - 3$

(3) $y = -x^2 - 2x + 2$

(4) $y = -\frac{1}{2}x^2 + 1$

(5) $y - 3 = 2(x - 2)^2$

#5 Identify the equations of the functions whose graphs are translated from the graph of

$y = \dfrac{1}{2}x^2$ in the following ways

(1) Translated -2 units along the x-axis

(2) Translated 2 units along the y-axis

(3) Translated 1 unit along the x-axis and -1 unit along the y-axis

(4) Translated -3 units along the x-axis and -4 units along the y-axis

(5) Translated m units along the x-axis and n units along the y-axis

#6 Find the value of a for which

(1) The graph of $y = ax^2$ passes through one point $(2, -2)$.

(2) The graph of $y = \left(x - \dfrac{1}{2}\right)^2 + a$ passes through one point $(-1, 3)$.

(3) The graph of $y = \left(x + \dfrac{a}{2}\right)^2 + 3$ has the vertex $(-5, 3)$.

(4) The graph of $y = \left(x - \dfrac{2a}{3}\right)^2 - 2$ has been translated from the graph of $y = x^2 - 2$,

  -4 units along the x-axis .

(5) The graph of $y = 2(x + 3a - 1)^2 + 1$ has the y-axis as the axis of symmetry.

#7 Find the value of ab when the parabola $y = -ax^2 + b$

(1) Passes through $(-1, 2)$ and $(3, -2)$

(2) Passes through $(1, 2)$ and $(-2, -4)$

#8 Find the value of $a + b$ for the following graphs of quadratic functions

(1) $y = 2x^2 - x + 3$ passes through the two points $(1, a)$ and $(-2, -b)$.

(2) $y = -ax^2 + 2x - 1$ passes through the two points $(1, b)$ and $(-1, a)$.

(3) $y = -x^2 + 2ax + 3$ passes through the two points $(-1, 0)$ and $(2, b)$.

#9 For any constants m, n, the parabola $y = \frac{1}{2}(x - m)^2 + n$ is translated from $y = \frac{1}{2}x^2$.

Give the conditions for m and n for the following parabola

(1)

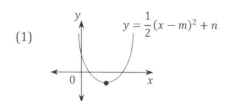

$y = \frac{1}{2}(x - m)^2 + n$

(2)

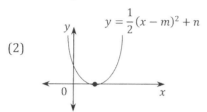

$y = \frac{1}{2}(x - m)^2 + n$

(3)

$y = \frac{1}{2}(x - m)^2 + n$

(4)

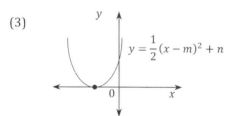

$y = \frac{1}{2}(x - m)^2 + n$

(5)

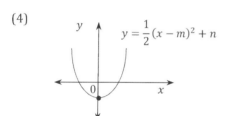

$y = \frac{1}{2}(x - m)^2 + n$

(6)

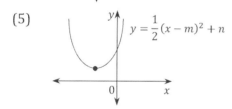

$y = \frac{1}{2}(x - m)^2 + n$

(7)

$y = \frac{1}{2}(x - m)^2 + n$

#10 Find an equation for the resulting quadratic function when

(1) The parabola $y = 2x^2 - 6x + 5$ is translated 2 units along the x-axis and -1 unit along the y-axis.

(2) The parabola $y = -3x^2 + 2x - 2$ is translated -2 units along the x-axis and 3 units along the y-axis.

(3) The parabola $y = -\frac{1}{2}(x + 2)^2 - 1$ is a symmetrical transformation along the x-axis.

(4) The parabola $y = \frac{1}{2}(x + 2)^2 + 1$ is a symmetrical transformation along the y-axis.

#11 Find the equation of the parabolas A and B on the graph.

Both are transformations of the parabola $y = \frac{1}{2}x^2$.

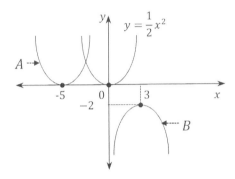

#12 State how the following parabolas have been translated from $y = -x^2 + 3x - 2$.

(1) $y = -x^2 + 4x + 3$

(2) $y = -x^2 + \frac{1}{2}x - 1$

(3) $y = -x^2 + 4x$

(4) $y = -x^2 + 3x + 2$

#13 Find the vertex, axis of symmetry, and intercepts for the following quadratic functions

(1) $y = 2x^2 + 3x + 1$

(2) $y = -x^2 + 2x + 3$

(3) $y = -3x^2 - 3x$

(4) $y = \frac{1}{2}x^2 - 4x + 6$

#14   Find the value of $a + p + q$ for the following quadratic functions

(1) $y = -3x^2 + 4x - a + 1 = a(x + p)^2 + q$

(2) $y = \frac{1}{2}x^2 - ax + 1 = \frac{1}{2}(x + 2)^2 + p + q$

(3) $y = ax^2 - 2x + 3 = -2(x + p)^2 - q$

#15   Find an equation of the quadratic function with following condition

(1) Vertex: $(1, 2)$ and passes through a point $(0, 3)$

(2) Axis of symmetry: $x = -1$ and passes through two points $(-3, -2)$, $(0, 4)$

(3) Vertex is on the x-axis, axis of symmetry: $x = -1$, and passes through a point $(-3, -4)$

(4) Passes through three points $(0, -3), (2, -1)$, and $(4, -6)$

(5) Passes through the origin, $(4, -3)$, and $(-2, 6)$

(6) Passes through $(-3, 0), (6, 0)$, and $(0, -6)$

(7) Passes through $(-4, 1), (-2, 0)$, and $(0, 3)$

#16   Find equations for the following parabolas

(1)

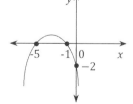

(2)

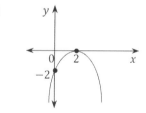

(3)

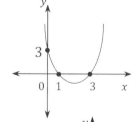

(4)

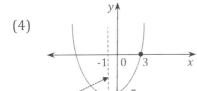

(5)

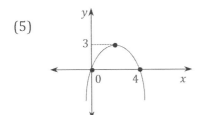

#17 Find the value of a for the following parabola

(1) The parabola $y = x^2 - ax + 2$ has $x = -2$ as its axis of symmetry.

(2) The parabola $y = -\frac{1}{2}x^2 + 4x - a + 1$ has its vertex on the x-axis.

(3) The distance between the two x-intercepts is 6 for a parabola $y = x^2 - 2x + a$.

#18 Find the minimum value or maximum values for the following quadratic functions

(1) $y = x^2 - 4x + 5$

(2) $y = -2x^2 + 4x + 1$

(3) $y = -3(x + 1)(x - 3)$

(4) $y = -x^2 + 4x - 4$

#19 Find the equation of the quadratic function with the following conditions

(1) The minimum value is 3 at $x = 1$ and passes through $(-2, 5)$.

(2) The maximum value is 4 at $x = -1$ and passes through $(1, -8)$.

#20 Find the value of $a + b$ for the following quadratic function with maximum or minimum values

    (1) $y = -x^2 + 2ax + b$ has the maximum value 4 at $x = 1$.

    (2) $y = 2x^2 - ax + b$ has the minimum value -3 at $x = -2$.

    (3) $y = ax^2 + 2x + b$ has the maximum value 3 at $x = 2$.

    (4) $y = ax^2 - bx + 2$ has the maximum value 3 at $x = -1$.

#21 One side of a rectangle is x inches. The perimeter and the area of a rectangle are 10 inches and y square inches, respectively. Find the maximum value of y.

#22 The sum of two numbers is 18. Find the maximum value of their product.

#23 The difference between two numbers is 10. Find the minimum value of their product.

#24 A ball is thrown upward from the top of a 5 feet table. After x seconds, the height of the ball from the ground is $y = -3x^2 + 12x + 5$. Find the maximum height from the ground the ball can reach.

Chapter16. Basic Statistical Graphs

16-1 Categorical Data

Providing information in graphic forms makes a stronger visual impact. So, using graphs is often helpful in understanding data.

1. Bar Graphs

A *bar graph* displays frequencies as does a frequency histogram. The difference between a bar graph and a frequency histogram is that a bar graph places class limits on the horizontal axis.

Example   Consider the following data:

66, 53, 41, 44, 63, 57, 61, 67, 72, 62

Class is a set of intervals (a range of values of a variable) into which the sample measurements may be grouped.

A histogram represents numbers by area, not height.

Class	Class limit	Measurements
40-50	41-50	41, 44
50-60	51-60	53, 57
60-70	61-70	61, 62, 63, 66, 67
70-80	71-80	72

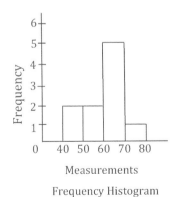

Frequency Histogram

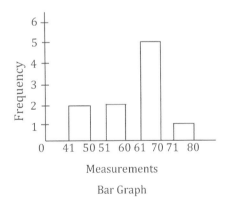

Bar Graph

We can place bars on a bar graph without any particular order because we don't place the class boundaries on the horizontal axis. Also, bar graphs can be used to represent two or more categorical measurements simultaneously. The height of each bar is proportional to the measurement value it represents.

2. Circle Graphs (Pie Charts)

Since a *circle graph* consists of parts, sectors, or sliced pieces in a circular pie, we can easily compare parts to each other and to the whole. By measuring the angle needed for each sector, we can draw the sectors correctly. The circle graph represents what percentage of the whole falls into each part.

Example   Consider the following distribution of mid-term grades in a math class.

Grade	Number of students
A	3
B	8
C	5
D	1
F	1

The total number of students is $3 + 8 + 5 + 1 + 1 = 18$.

Now, find the angle needed for each sector.

Grade A : $\frac{3}{18} \times 360° = 60°$

Grade B : $\frac{8}{18} \times 360° = 160°$

Grade C : $\frac{5}{18} \times 360° = 100°$

The sum of the angle measure for all the sectors in a circle is 360°.

Grade D : $\frac{1}{18} \times 360° = 20°$

Grade F : $\frac{1}{18} \times 360° = 20°$

Using a protractor to measure angles, we can draw a precise circle graph.

Distribution of Grades

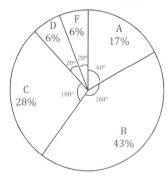

We can more easily obtain information about the relative size of parts with a circle graph rather than with a table.

16-2 Measurement Variables

These help to detect the pattern (trend)!

1. Line Graphs

A *line graph* provides the same information as a histogram. Instead of using bars, we place a point at the correct height for each class interval and connect all the points with line segments. A steeper line segment represents that more change has occurred during that class interval. Observing a line graph, we can detect an overall pattern over a measurement variable so that we can make predictions.

Example   Make a line graph for the following data:

Number of books checked out from a library

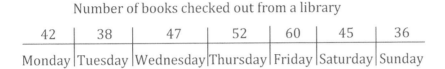

42	38	47	52	60	45	36
Monday	Tuesday	Wednesday	Thursday	Friday	Saturday	Sunday

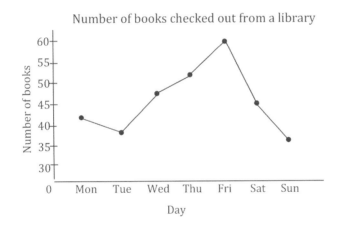

2. Scatter Plots

Even though a scatter plot can be more complicated to understand than a line graph, it represents more information about 3 important elements : outliers, the extent of variability for a measurement variable at each location of the other measurement variable, and an overall pattern for a measurement variable to increase as the other measurement variable increases. Thus, *scatter plots* are the most useful graphs for the relationship between two measurement variables. If variables are clearly related, a scatter plot can highlight the correlation between them.

A scatter plot is a plot of x_i and y_i as points (x_i, y_i) in a coordinate plane, where x_i and y_i are the values of measurement variables x and y for the same object.

For example, consider the following scatter plot displaying the relationship between the performance scores and daily practice minutes of every student in a class:

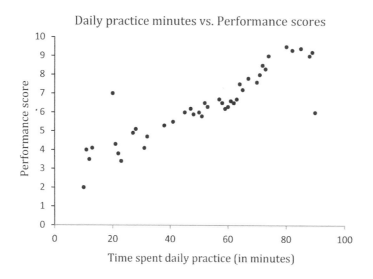

Each "•" on the plot represents a student.

The scatter plot shows an increasing trend toward higher scores with longer periods of practicing and also shows that there is still substantial variability in performance scores at each period of time spent in daily practice. Moreover, there are a few outliers. One student who only spent 20 minutes still earned 7 points in his/her performance, another student spent 90 minutes practicing and only earned 6 points in his/her performance, the same score as several students who spent a significantly shorter period of time practicing.

Exercises

#1 Refer to the bar graph below to answer the following questions.

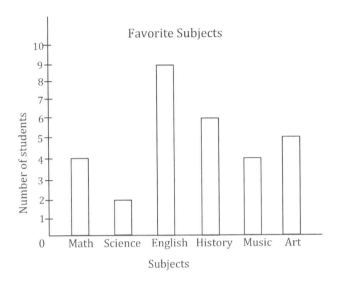

(1) Which subjects are the most and least favorite subjects?

(2) Which two subjects are chosen by the same number of students?

(3) How many more students chose English than Math?

(4) What percentage of students chose history?

#2 Present the following information in a circle graph.

Item	Price
Shoes	$45
Pants	$30
Food	$20
Snack	$15
Bag	$10

#3 According to the circle graph, how much did Nichole spend on each item for the current month?

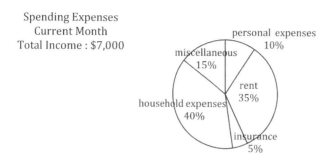

Spending Expenses
Current Month
Total Income : $7,000

#4 Complete the following table using the categorical data listed in the table:

Category	Frequency	Relative Frequency	Central Angle (in degree)
A	50		
B	85		
C	65		
D	40		
Total	240	1.000	360

#5 Refer to the line graph below to answer the following questions.

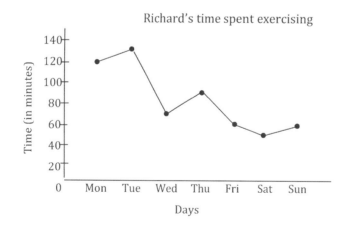

Richard's time spent exercising

(1) On which day did Richard exercise most?

(2) Which days did Richard exercise for the same amount of time?

(3) Which two days show the greatest difference between Richard's time spent exercising?

Chapter 17. Descriptive Statistics

Statistics is the branch of scientific inquiry that provides methods for organizing and analyzing data. Studying statistics helps us to obtain an understanding from numbers. The study of statistics also includes the study of the development of techniques for collecting data.

17-1   Organizing Data

1. Stem Plots (Stem-and-Leaf Plots)

A *stem plot* is a convenient way to put a list of measurements into order while getting its shape. Stem plots provide a graphic description of summarizing data.

To make a stem plot for data,

(1) Separate each observation into a "stem" and "leaf". Generally, stems may have as many digits as needed, but each leaf should contain only one digit.

(2) List the stems in a vertical column from the smallest to the largest. Place a vertical line to the right of the stems, and add the leaves in the display row which corresponds to the observation's stem (from the smallest to the largest).

Using a stem plot, we can identify the center (median), which divides the data set into two sets of equal size. We can also identify the overall shape (symmetric or skewed in one direction) of the distribution. After observing the overall shape, we can check if outliers, or individual values that stand apart from the usual pattern, exist.

Example

Consider the test scores of room 6 in a school.

90, 75, 82, 69, 77, 88, 88, 86, 90, 100, 84

List all the tens digits of each number as stems and the ones digits as leaves.

The resulting stem plot looks like this

```
 6 | 9
 7 | 5 7
 8 | 2 4 6 8 8
 9 | 0 0
10 | 0
```

We find the center (median) of the distribution by counting the same amount of numbers from each end of the stem plot. That is, the center (median) is 86. The stem plot also shows that the distribution is symmetrical around the middle value.

6 | 9   means 69

The stem plot is the simplest type of graphic description for a small number of observations in a data set. To create a graphic description of a large number of data sets, we use a relative frequency histogram which doesn't list all data values.

2. Relative Frequency Histogram

Frequency distribution provides a more compact summary of a data set than does a stem plot display.

A graphical representation of a frequency distribution can be obtained by constructing a histogram.

When frequencies are represented along the vertical axis, a histogram is called a frequency histogram.

A large number of observations on a single variable can be summarized in a table of frequencies or relative frequencies.

To make a frequency histogram,

(1) Divide the *range* (the interval between the greatest and smallest measurement) into *classes* (equal subintervals). These classes must be specified precisely so that each measurement falls in to exactly one class.

(2) Count how many measurements fall into each class of the range.

> *Note*   *For the frequencies* f_1, f_2, f_3, $\cdots\cdots$,
>
> *the quantities* $\dfrac{f_1}{n}$, $\dfrac{f_2}{n}$, $\dfrac{f_3}{n}$, $\cdots\cdots$, *where* n *is the total number of frequency,*
>
> *are called relative frequencies.*
>
> *Relative frequency is the proportion of observations falling in to the corresponding class interval.*

(3) Draw a horizontal line to represent the measurement axis and a vertical axis to represent the frequency scale.

(4) Above each class interval, draw a bar with a height equal to the count for each class of the range. The heights of the bars are represented as proportions.

Example

A frequency distribution for test scores

Class Interval (Test Scores)	Frequency (Number of Students)	Relative Frequency (Number of Students)
65-70	1	$\frac{1}{50}$
70-75	4	$\frac{4}{50}$
75-80	12	$\frac{12}{50} = \frac{6}{25}$
80-85	18	$\frac{18}{50} = \frac{9}{25}$
85-90	8	$\frac{8}{50} = \frac{4}{25}$
90-95	5	$\frac{5}{50} = \frac{1}{10}$
95-100	2	$\frac{2}{50} = \frac{1}{25}$
Total	50	

Frequency histogram for test scores

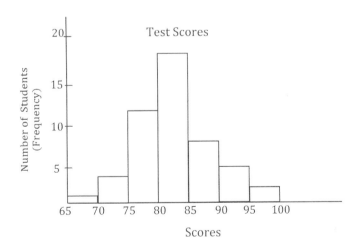

The distribution is centered in the 81 to 85 class. It is roughly symmetrical in shape.

The histogram shows that there are no large gaps or obvious outliers.

17-2 Measures of Central Tendency

Identifying the numerical measure of the center of a distribution is helpful when inspecting a stem plot or a histogram.

The most common numerical measure for describing the location of a data set is the arithmetic mean (or, also referred to simply as the mean).

1. The Mean

The most measure of central tendency (the tendency of the data to center about certain numerical values) for a quantitative data set is the mean of the data set.

The *mean* of a set of quantitative data is equal to the sum of all the sample measurements divided by the total number of measurements in the sample.

That is, for a given set of numbers $x_1, x_2, x_3, \cdots\cdots, x_n$, the mean is denoted by

$$\bar{x} = \frac{x_1 + x_2 + x_3 \cdots\cdots + x_n}{n}.$$

Example   Consider the data set: 1, 2, 3, 4, 5, 6, 9, 10

Then, the mean is

$$\bar{x} = \frac{1+2+3+4+5+6+9+10}{8} = \frac{40}{8} = 5.$$

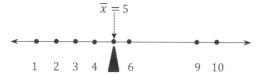

The mean also is the balance point for a system of weights.

Since the mean $\bar{x}$ represents the average value of the measurements in a sample data set, we can find the average, denoted by μ, of all values in the population.

The *population mean μ* is defined by

$$\mu = \frac{\text{Sum of } N \text{ population values}}{N}, \text{ for } N \text{ population.}$$

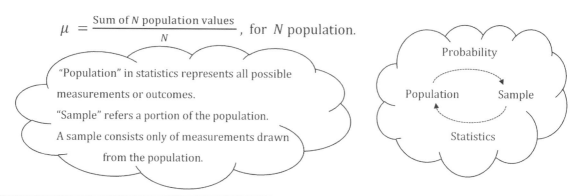

"Population" in statistics represents all possible measurements or outcomes.

"Sample" refers a portion of the population.

A sample consists only of measurements drawn from the population.

Probability

Population   Sample

Statistics

2. The Median

The mean as a measure of center is sensitive to the influence of the few extreme observations, also called outliers.

The median is a measure that is less sensitive to outlying values than the mean $\overline{x}$.

The *median* is the middle number of a distribution when the measurements in a data set are arranged in ascending order (from smallest value to largest value), including all repeated values.

If the number n of total measurements is odd, then the median M is the middle number in the ordered list. That is $M = \left(\frac{n+1}{2}\right)^{th}$ value in the ordered list.

If the number n of total measurements is even, then the median M is the mean of the two middle measurements in the ordered list.

That is M is equal to the mean of the $\left(\frac{n}{2}\right)^{th}$ and $\left(\frac{n}{2} + 1\right)^{th}$ ordered values.

Example

Consider the following data

$$15,\ 9,\ 7,\ 11,\ 8,\ 20,\ 11,\ 14,\ 9$$

To find the median, arrange all measurements in order from smallest to largest including repeated values. Then, the list of ordered values is

Order	1th	2nd	3rd	4th	5th	6th	7th	8th	9th
Measurement	7	8	9	9	(11)	11	14	15	20

Median

Since the number n of total measurements is $n = 9$ (odd number),

the median is the middle value, 11 (the $\left(\frac{n+1}{2}\right)^{th}$ value $= \left(\frac{9+1}{2}\right)^{th}$ value $= 5^{th}$ value).

If the largest measurement (20) had not appeared in this data, then $n = 8$ (even number).

Order	1th	2nd	3rd	4th	5th	6th	7th	8th
Measurement	7	8	9	9	11	11	14	15

$$\text{Median} = \frac{9+11}{2} = 10$$

So, the median is the mean of the 4th value and the 5th value.

Therefore, the median is $M = \frac{9+11}{2} = \frac{20}{2} = 10$.

3. The Mode

The *mode* is the measurement which occurs with the greatest frequency in a set of data.
The mode is of main value in describing large data set. The mode in a large data set will be in the class containing the largest relative frequency in a relative frequency histogram. Thus, the mode of a set of measurements is a useful measure of central tendency.

Example   Consider the following data

$$7,\ 2,\ 3,\ 3,\ 5,\ 3,\ 1,\ 2,\ 4,\ 2$$

Since 2 and 3 each occur three times (∵ more often than any other value), the modes are 2 and 3. On the other hand, the data might consist of values, each of which occurs the same number of times.

Example   Consider the following data sets of A and B

$$A:\ 3,\ 7,\ 5,\ 1,\ 4$$
$$B:\ 6,\ 5,\ 6,\ 3,\ 5,\ 3$$

The two data sets have no mode. Therefore, for a given set of data, it is possible to have more than one mode or no mode at all.

4. Comparing the Mean, Median, and Mode

The mean, median, and mode are all measures for the center of a data set. But, the mean and median focus on different aspects of the sample data. So, they are not equal generally.

For a population mean and a population median,

① **Mode $<$ Median $<$ Mean**

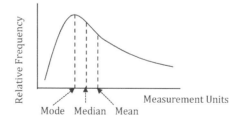

The data set is skewed to the right.
(Positively skewed)
The higher values are more spread out than the lower values.

Note   A positive skew indicates that the tail on the right side is longer than the left side and the bulk of the values lie to the left of the mean.

② **Mean = Median = Mode**

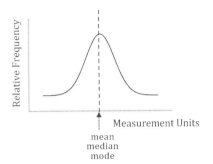

The data set is symmetrical

Mirror image of the shape

  on the other side about the center line

Bell curve data set

③ **Mean < Median < Mode**

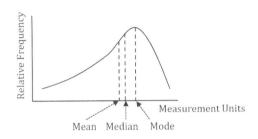

The data set is skewed to the left.

(Negatively skewed)

The lower values are more spread out

and the higher values tend to be

clumped.

Note   A negative skew indicates that the tail on the left side is longer than the right side and the

bulk of the values lie to the right of the mean.

A skewed data set is substantially

off from being a bell curve.

17-3 Measures of Variability (Spread)

1. The Range

Consider the following two histograms

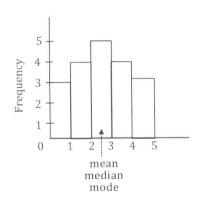

Figure 1

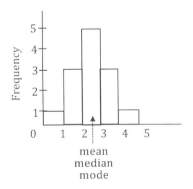

Figure 2

From the two histograms, we notice that both sets of data are symmetrical with equal means, medians, and modes. But Figure 1 and Figure 2 differ substantially in the variability or spread of the data set. Figure 1 shows an almost equal relative frequency, while most of the measurements in Figure 2 are clustered around their center. That is, the measurements in Figure 1 are more spread out or highly variable than the measurements in Figure 2 and the measurements in Figure 2 do not display much variability.

Therefore, we need to identify both a measure of relative variability as well as a measure of central tendency to describe these data sets.

The simplest measure of variability of a data set is its range.

The *range* of a data set is the arithmetic difference between the largest measurement and the smallest measurement.

Note: The ranges of the data sets in Figure 1 and Figure 2 are 5 – 3 = 2 and 5 – 1 = 4, respectively. Thus, Figure 2 shows less variability than Figure 1.

Example

Consider the following data

72, 64, 66, 75, 59

Since the largest measurement of the data is 75 and the smallest measurement is 59, the range of the data is $75 - 59 = 16$.

2. Variance and Standard Deviation

Consider the following two sets of data A and B.

A : 5, 8, 11, 25, 26, 27, 30, 45, 55, 68

B : 5, 6, 7, 8, 10, 30, 43, 56, 67, 68

Since A and B have the same mean, 30, and range, 63, the two measures of central tendency are not distinguished between the two data sets. However, they have a distinctly different variability. The measurements in data set A cluster more closely around their center than do the measurements in data set B. Concerning an effective measure of variability, we consider the deviation and variance.

For any measurements x_1, x_2, x_3, $\cdots\cdots$, x_n,

$x_i - \overline{x}$ is called the *deviation* of the ith measurement from the mean, $\overline{x}$.

However, $(x_1 - \overline{x}) + (x_2 - \overline{x}) + \cdots\cdots + (x_n - \overline{x})$

$$= (x_1 + x_2 + \cdots + x_n) - n \cdot \overline{x}$$

$$= (x_1 + x_2 + \cdots + x_n) - n \cdot \frac{(x_1+x_2+\cdots+x_n)}{n} = 0$$

That means, the average deviation from the mean is always 0.

Thus, we consider $(x_i - \overline{x})^2$ instead of $(x_i - \overline{x})$.

For any measurements x_1, x_2, x_3, $\cdots\cdots$, x_n, the *variance* of a set of data, denoted by S^2, is defined by

$$S^2 = \frac{(x_1-\overline{x})^2+(x_2-\overline{x})^2+\cdots\cdots+(x_n-\overline{x})^2}{n} = \frac{\sum\limits_{i=1}^{n}(x_i-\overline{x})^2}{n}$$

> The standard deviation measures a set of data spreads out around its average. So, we can find how far away numbers on a list are from their average.

The *standard deviation* of a set of data, denoted by s,

is the positive square root of the variance S^2.

That is, $\boxed{s = \sqrt{S^2}}$

Variance and standard deviation are the most useful measures of variability.

If there are N values in the finite population, then the population mean is

$$\mu = \frac{\text{Sum of the } N \text{ population values}}{N}$$

(Note that: $\overline{x}$ represents the mean value of the observations in a sample.)

When the population is finite and consists of N values, then

$$S^2 = \frac{\sum\limits_{i=1}^{N}(x_i-\overline{x})^2}{N} \quad ; \text{ The average of all squared deviations from the population mean}$$

(For the population, the divisor is N, not $N-1$.)

Since the value of μ is almost never known,

the sum of squared deviations about $\overline{x}$ must be used.

Since the $x_i's$ tend to be closer to their average $\overline{x}$ than to the population average μ,

the divisor $n-1$ is used rather than n.

If you use a divisor n in the sample variance, the result would tend to underestimate S^2

(produce too small estimated values on the average), while dividing by $n-1$ corrects the

underestimating.

Note that
$$S^2 = \frac{\sum_{i=1}^{n}(x_i - \overline{x})^2}{n} = \frac{1}{n}\sum_{i=1}^{n}x_i^2 - (\overline{x})^2$$

The standard deviation of a set is a measure of how much a number in the set differs from the mean. The greater the standard deviation, the more the numbers in the set vary from the mean.

Example   Consider the three sets with means of 5
$$A:\{5,5,5,5\}, \quad B = \{3,3,7,7\}, \quad C = \{3,4,5,8\}$$

The standard deviations of the sets are

$$S_A = \sqrt{\frac{(5-5)^2+(5-5)^2+(5-5)^2+(5-5)^2}{4}} = 0$$

$$S_B = \sqrt{\frac{(5-3)^2+(5-3)^2+(5-7)^2+(5-7)^2}{4}} = \sqrt{\frac{16}{4}} = 2$$

$$S_C = \sqrt{\frac{(5-3)^2+(5-4)^2+(5-5)^2+(5-8)^2}{4}} = \sqrt{\frac{4+1+0+9}{4}} = \sqrt{\frac{14}{4}} = \sqrt{7}$$

Example   Consider the following data
$$2, \ 3, \ 4, \ 3, \ 5, \ 7$$

To obtain the variance and standard deviation, determine the mean, $\overline{x}$.

Since $\overline{x} = \frac{2+3+4+3+5+7}{6} = \frac{24}{6} = 4$,

the variance S^2 is

$$S^2 = \frac{(2-4)^2+(3-4)^2+(4-4)^2+(3-4)^2+(5-4)^2+(7-4)^2}{6} = \frac{4+1+0+1+1+9}{6} = \frac{16}{6} \approx 2.67$$

and the standard deviation is $s = \sqrt{S^2} \approx \sqrt{2.67}$

Variance $= \dfrac{\text{Sum of squared deviation}}{n}$

Standard deviation $= \sqrt{\text{Variance}}$

Example   Consider a set of data: 3, 4, 4, 5, 5, 7, 7, 7, 8, 9

Find the standard deviation of the data.

$$\overline{x} = \frac{59}{10} = 5.9$$

$$s = \sqrt{\frac{1}{n}\sum x_i^2 - (\overline{x})^2} = \sqrt{\frac{3^2+2\cdot4^2+2\cdot5^2+2\cdot5^2+3\cdot7^2+8^2+9^2}{10} - (5.9)^2} = \sqrt{\frac{383}{10} - 34.81}$$

$$= \sqrt{3.49} \approx 1.87$$

Note that $s = \sqrt{\dfrac{\sum(x_i - \bar{x})^2}{n}} = \sqrt{\dfrac{(2.9)^2 + 2(1.9)^2 + 2(0.9)^2 + 3(1.1)^2 + (2.1)^2 + (3.1)^2}{10}}$

$\qquad = \sqrt{\dfrac{8.41 + 7.22 + 1.62 + 3.63 + 4.41 + 9.61}{10}} = \sqrt{\dfrac{34.9}{10}} = \sqrt{3.49} \approx 1.87$

Therefore, $s = \sqrt{\dfrac{\sum(x_i - \bar{x})^2}{n}} = \sqrt{\dfrac{1}{n}\sum x_i^2 - (\bar{x})^2}$

3. Quartiles

The variability or spread of a distribution can also be indicated by percentile.

Note   *The median is the 50th percentile.*

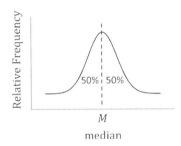

The most commonly used percentiles other than the median are the quartiles.

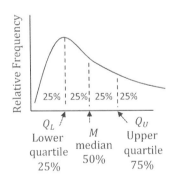

The quartiles divide a data set into four groups, each containing 25% of the measurements. The *lower quartile*, Q_L is the 25th percentile (the median of the smallest half of the data), the *middle quartile* is the 50th percentile (the median M), and the *upper quartile*, Q_U is the 75th percentile (the median of the largest half of the data).

Example

Consider the following data

1, 2, 2, 3, 4, 5, 5, 6, 7

The data set is arranged in order and the number n of measurements is $n = 9$ (odd).

So, the median M is the middle value $\left(\frac{n+1}{2} = \frac{9+1}{2} = 5^{th} \right)$, 4.

Since the median M divides the data set into two equal halves, the lower quartile is the median of the lower half : 1, 2, 2, 3 and the upper quartile is the median of the upper half : 5, 5, 6, 7. Since each half contains $n = 4$ (even), of measurements, the median is the mean of two middle values. That is, the lower quartile is $\frac{2+2}{2} = 2$ and the upper quartile is $\frac{5+6}{2} = 5.5$

> Quartiles are found in ascending order.

4. Boxplots (Box-and-Whisker Plots)

Boxplots provide a method for detecting outliers (measurements that are further away from the rest of the data) based on the distance between the lower quartile, Q_L and the upper quartile, Q_U. *Interquartile range* (IR) is defined by $IR = Q_U - Q_L$.

Boxplots show the center (median), variability (spread), the extent of skewness from symmetry, and the detection of outliers.

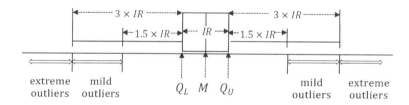

Example   Consider the following data

0, 1, 5, 5, 7, 8, 9, 10, 24, 33

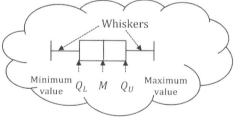

Since the sample size is $n = 10$ (even), the median M is the mean of $\left(\frac{n}{2} \right)^{th}$ and $\left(\frac{n}{2} + 1 \right)^{th}$ values. That is, $M = \frac{7+8}{2} = 7.5$ The lower quartile Q_L is 5 and the upper quartile Q_U is 10.

Order	1th	2nd	3rd	4th	5th	6th	7th	8th	9th	10th
Measurement	1	1	(5)	5	7	8	9	(10)	24	33
			Q_L			M		Q_U		

So, $IR = Q_U - Q_L = 10 - 5 = 5$.

Since $1.5 \times IR = 7.5$ and $3 \times IR = 15$, the boxplot is :

$Q_L = 5$ $M = 7.5$ $Q_U = 10$ 17.5   25

From the boxplot, we see

① The median is in the middle of the box.

② The middle 50% (from Q_L to Q_U) of the data set shows that the data forms a symmetrical distribution.

③ There is a mild outlier (indicated by open circle), 24, and an extreme outlier (indicated by closed circle), 33.

5. Line Plots

If a set of data has distinct measurements, we use a line plot with an $\times$ for each measurement whenever it appears in the set of data. Line plots display the frequencies in the data set, as does a frequency histogram.

Using a line plot, we can also obtain the mode, median, range, and outliers.

Example

Consider the following frequency distribution of measurements.

Measurement	Frequency	Relative Frequency
1	4	$\frac{4}{20} = \frac{1}{5} = 20\%$
2	2	$\frac{2}{20} = \frac{1}{10} = 10\%$
3	6	$\frac{6}{20} = \frac{3}{10} = 30\%$
4	4	$\frac{4}{20} = \frac{1}{5} = 20\%$
5	2	$\frac{2}{20} = \frac{1}{10} = 10\%$
6	2	$\frac{2}{20} = \frac{1}{10} = 10\%$

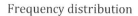

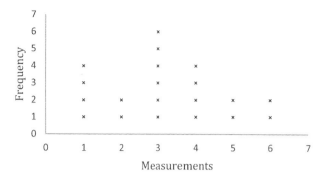

From the line plot, we observe that the mode is 3, the median is 3, and the range is 6−1=5.

The mean is $\dfrac{1 \cdot 4 + 2 \cdot 2 + 3 \cdot 6 + 4 \cdot 4 + 5 \cdot 2 + 6 \cdot 2}{20} = \dfrac{64}{20} = 3.2$

Exercises

#1 Make a stem-and-leaf plot for the following data.

43, 100, 50, 64, 73, 79, 81, 66, 55, 61, 101, 52, 55, 48, 64, 113, 77, 80, 81, 95, 53

#2 The following are the hourly wages in dollars of 20 workers. Arrange the data given below in a frequency table.

9.80 9.60 10.15 9.80 10.60 12.20 8.85 11.50 9.60 10.20

10.15 8.85 9.80 10.20 10.15 9.80 11.50 9.80 9.80 9.60

#3 The following distribution shows the number of days in a year each student in a class of 50 visited a doctor. Draw a relative frequency histogram.

Number of days	Number of students
1-5	4
6-10	6
11-15	11
16-20	13
21-25	9
26-30	7

#4 Calculate the mean, median, mode, and range for the following sample measurements

(1) 3, 6, 7, 5, 4, 3, 2

(2) 9, 3, 5, 5, 2, 20, 4, 6

(3) 23.5, 31.2, 18.4, 35.4, 25

(4) −4, −5, −7, −3, −4, −7, −2, −6

#5 Find the deviations, variance, and standard deviation for the following data

25, 30, 36, 38, 43, 56

#6 Refer to the data set below to answer the following questions.

27, 26, 27, 38, 23, 27, 39, 42, 38, 63, 34

(1) Determine the lower quartile, Q_L and upper quartile, Q_U .

(2) Calculate the value of the interquartile range (IR).

(3) Draw a box plot.

#7 Complete the following table and draw a line plot for the frequency distribution.

Measurement	Frequency	Relative Frequency (%)
36.4	3	
42.5	6	
48.1	4	
53.2	6	
55.8	8	
64.7	9	
66.3	6	
72.9	3	
total	45	$1 = 100\%$

#8 Refer to the data set below to answer the following questions.

Age of teachers in a certain school

43 40 55 47 37 36 52 40 28 26 42 40 36 45 39

(1) Draw a stem-and-leaf plot for the data.

(2) Complete the frequency table.

Age	Frequency	Relative Frequency (%)
26-30		
31-35		
36-40		
41-45		
46-50		
51-55		

(3) Draw a relative frequency histogram.

(4) Find the mean, median, mode, and range.

(5) Draw a line plot.

(6) Draw a box plot.

Chapter 18. The Concept of Sets

18-1 Sets

1. Definition

A *set* is any collection into a whole of definite, distinguishable objects, called *elements*.

For example, ① The set of all books in a library.

② The set of all students in a classroom.

③ The set of letters a, b, and c.

④ The set of rules in a group.

⑤ The set of all positive integers whose square is 2.

⑥ The set of all numbers greater than 3.

(1) Finite Sets

A finite set is a set which contains only a finite number of elements.

Examples ① to ⑤ above are all finite sets.

(2) Infinite Sets

An infinite set is a set which contains an infinite number of elements

Example ⑥ above is an infinite set.

(3) Empty Set

An empty set is a set which has no elements.

Example ⑤ above is an empty set.

> Empty set , ∅, has no element.
> ∅ has 0 element.
> ∅ is a finite set.

2. Symbols

Sets are designated by the following symbols.

(1) { }   The symbol { } represents the elements in a set.

The set example ③ above is $\{a, b, c\}$, and the set example ⑥ above is $\{4, 5, 6, \cdots \cdots \}$.

(2) ∅   An empty set is denoted by the symbol ∅.

The set example ⑤ above is ∅.

(3) ∈ , ∉   The symbols ∈ , ∉ designate whether an element belongs (or doesn't belong) to a set.

① If a is an element of the set A, then we write $a \in A$ (a belongs to A.).

② If a is not an element of the set A, then we write $a \notin A$ (a does not belong to A.).

For example, $A = \{1, 2, 3\} \Rightarrow 3 \in A, \ 4 \notin A$

(4) $n(A)$ The number of elements in a set A is denoted by $n(A)$.

For example, ① If A is the set of all prime numbers less than 10, then

$$n(A) = n(\{2, 3, 5, 7\}) = 4$$

② If $A = \emptyset$, then $n(A) = n(\emptyset) = 0$

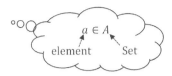

3. Set Notation

(1) A set $A = \{a, b, c\}$ is the same as $\{b, a, c\}$ or $\{c, b, a\}$, etc.

Note ① $\{a, a, b\}$ *is not a proper notation of a set, because all the elements in a set are distinct.*
It should be replaced by $\{a, b\}$.

② $\{a, b, c, \cdots\cdots\}$ *is used when a set includes many elements and there is a fixed rule between the*
elements. For example, a set of even numbers is $\{2, 4, 6, \cdots\cdots\}$.

(2) The set builder notation $\{x \in A \mid P(x)\}$

$\{x \in A \mid P(x)\}$ is the set of all x in A such that $P(x)$ is true.

As a rule, to every set A and to every statement $P(x)$ about $x \in A$, there is a set

$\{x \in A \mid P(x)\}$ whose elements are precisely those elements x of A for which the statement

$P(x)$ is true.

Example

Let A be the set of all students in a class.

The statement "x is a boy." is true for some elements x of A and false for others.

To specify the set of all the boys in the class, we use the notation $\{x \in A \mid x$ is a boy. $\}$.

Similarly, $\{x \in A \mid x$ is not a boy. $\}$ specifies the set of all the girls in the class.

(3) Venn Diagram

Using a diagram called the Venn Diagram, the relationship between sets are visualized and
illustrated.

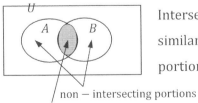

Intersecting portion of two circles A and B represents
similarities between two sets, while non-intersecting
portions of the circles represent differences between two sets.

non − intersecting portions

intersecting portion

18-2 Subsets

1. Definition

(1) Identical

Two sets are called *identical* if every element of each set is an element of the other set.

Identical sets are denoted by $A = B$.

Note If two sets A and B are equal or identical, then A = B (Both sets contain the same element).

(2) Subset

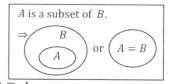

If all elements of a set A are elements of a set B,

then set A is called a *subset* of a set B and denoted by $A \subseteq B$ or $B \supseteq A$.

(3) Superset

If a set A is a subset of a set B, then B is called a *superset* of A.

(4) Proper Subset

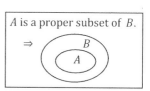

If $A \subseteq B$ and $A \neq B$,

then we write $A \subset B$ or $B \supset A$ and

A is called a *proper subset* of B or B is called a *proper superset* of A.

If A is not a subset of B, then we denote $A \nsubseteq B$.

2. Properties of Subsets

(1) The empty set $\emptyset$ is a subset of every set.

$\emptyset \subset A,\ \emptyset \subset \emptyset$

(2) Every set is a subset and a superset of itself.

$A \subseteq A$

(3) If $A \subseteq B$ and $B \subseteq C$, then $A \subseteq C$.

(4) If $A \subset B$ and $B \subset A$, then $A = B$.

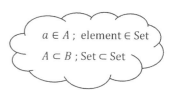

3. Numbers of Subsets

For any set A which has $n(A) = m$,

(1) The number of subsets is $2^m = \underbrace{2 \times 2 \times \cdots\cdots \times 2 \times 2}_{m \text{ times}}$

(2) The number of proper subsets is $2^m - 1$

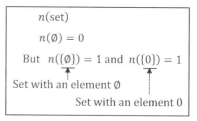

Example $A = \{a, b, c\}$

The subsets of A are $\emptyset, \; \{a\}, \; \{b\}, \; \{c\}, \; \{a, b\}, \; \{a, c\}, \; \{b, c\}, \; \{a, b, c\}.$

0-element   1-element   2-elements   3-elements

So, the set A has 8 subsets.

Since $n(A) = 3$, the number of subsets is $2^3 = 8$.

Since the proper subset does not include itself, the number of proper subsets is $2^3 - 1 = 7$.

Note ① $A = \emptyset \Rightarrow n(A) = 0 \;$ and $\; n(A) = 0 \Rightarrow A = \emptyset$

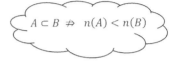

$A \subset B \not\Rightarrow n(A) < n(B)$

      ② $A \subset B \Rightarrow n(A) \leq n(B) \quad$ but, $n(A) \leq n(B) \not\Rightarrow A \subset B$

                 ($\because$ For example, $A = \{a, b\}, \; B = \{b, c, d\}$

                        $\Rightarrow n(A) = 2 < 3 = n(B) \;$ but $\; A \not\subset B$)

      ③ $A = B \Rightarrow n(A) = n(B) \quad$ but, $n(A) = n(B) \not\Rightarrow A = B$

(3) The number of subsets which do include (or do not include) certain elements is

     2^{m-p}**, where** $p = n(\{\text{cetain elements}\})$

Example $A = \{d, e, f\}$

① The number of subsets which include d :

To find the subsets which include the element "d", first consider the subsets of $\{e, f\}$ which exclude the element "d". Then, put the element "d" back in the subsets.

So, the number of subsets which include the element "d" is the number of subsets of $\{e, f\}$.

$2^{3-1} = 2^2 = 4$

② The number of subsets which do not include d :

Find the subsets excluding the element "d".

So, it will be the same as the number of subsets of $\{e, f\}$.

$2^{3-1} = 2^2 = 4$

③ The number of subsets which include d but not e:

It will be the same as the number of subsets of $\{f\}$.

$2^{3-1-1} = 2^1 = 2$

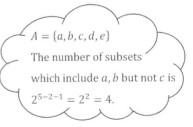

$A = \{a, b, c, d, e\}$

The number of subsets which include a, b but not c is

$2^{5-2-1} = 2^2 = 4$.

18-3 Operations on Sets

1. Unions and Intersections

(1) Definition

1) **Union**: The *union* of any two sets A and B is the set of all elements x such that x belongs to at least one of the two sets A and B . The union is denoted by $A \cup B$.

$$A \cup B = \{ x \in A \cup B \mid x \in A \text{ or } x \in B \}$$

Note: $A \cup B$ is shaded.

     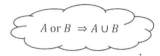

Figure 1   Figure 2   Figure 3

2) **Intersection**: The *intersection* of any two sets A and B is the set of all elements x which belong to both A and B . The intersection is denoted by $A \cap B$.

$$A \cap B = \{ x \in A \cap B \mid x \in A \text{ and } x \in B \}$$

Note: $A \cap B$ is shaded.

     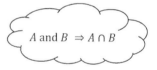

Figure 1   Figure 2   Figure 3

(2) The Property of Unions and Intersections

For any sets A and B,

1) $A \cup \emptyset = A, \qquad A \cap \emptyset = \emptyset$

2) $A \cup A = A, \qquad A \cap A = A$

3) $A \cup B = B \cup A, \qquad A \cap B = B \cap A$

4) $A \subset B \ \Rightarrow \ A \cup B = B, \ A \cap B = A$

5) $(A \cap B) \subset A \subset A \cup B, \qquad (A \cap B) \subset B \subset A \cup B$

(3) The Number of Elements

For any finite sets A and B,

1) $n(A \cup B) = n(A) + n(B) - n(A \cap B)$

2) $n(A \cap B) = n(A) + n(B) - n(A \cup B)$

Example

$A = \{a, b, c\}, \; B = \{b, c, d, e\}$

$\Rightarrow \; A \cup B = \{a, b, c, d, e\} \;$ and $\; A \cap B = \{b, c\}$

Using a Venn Diagram,

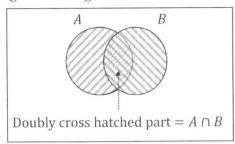

Doubly cross hatched part $= A \cap B$

$A \cap B$ belongs to A and also B. That means $n(A \cap B)$ is doubly added to $n(A) + n(B)$.

So, $n(A \cap B)$ should be subtracted one time from $n(A) + n(B)$.

$n(A \cup B) = n(A) + n(B) - n(A \cap B)$.

Note For any finite sets $A, B,$ and $C,$

$$n(A \cup B \cup C) = n(A) + n(B) + n(C) - n(A \cap B) - n(B \cap C) - n(A \cap C) + n(A \cap B \cap C).$$

2. Complements

The operation for complementation is similar to the operation of subtraction.

(1) Definition

1) A^C   For any subset A of the universal set $U,$ the set of all the elements which do not

belong to A is denoted by A^C. $A^C = \{x \mid A \subset U, \; x \in U, \; x \notin A\}$

Note: For a fixed set U which may be regarded as a universal set, $U - A = A^C$

2) $A - B$ The relative complement of a set B in a set A is the set $A - B$

and it is denoted by $A - B$. $A - B = \{x \mid x \in A \text{ and } x \notin B\}$

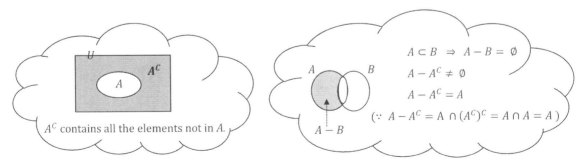

A^C contains all the elements not in A.

$A - B$

$A \subset B \; \Rightarrow \; A - B = \emptyset$

$A - A^C \neq \emptyset$

$A - A^C = A$

$(\because A - A^C = A \cap (A^C)^C = A \cap A = A)$

(2) Properties of Complements

For any subsets A and B of the universal set U,

1) ① $(A^C)^C = A$

② $\emptyset^C = U, \; U^C = \emptyset$

③ $A \cup A^C = U, \; A \cap A^C = \emptyset$

④ $A \subseteq B$ if and only if $B^C \subseteq A^C$

Note **De Morgan's theorem**

$$\text{For any two sets } A \text{ and } B, \quad \begin{cases} \text{①} \; (A \cup B)^C = A^C \cap B^C \\ \text{②} \; (A \cap B)^C = A^C \cup B^C \end{cases}$$

2) ① $U - A = A^C, \quad U - \emptyset = U$

② $A - A = \emptyset, \quad A - \emptyset = A$

③ $A - B = A \cap B^C = A - (A \cap B) = A \cup B - B$

④ $A - B = \emptyset \;\Rightarrow\; A \subseteq B$

⑤ $A \cap B = \emptyset \;\Rightarrow\; A - B = A, \; B - A = B$

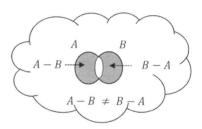

(3) The Number of Elements

For any subsets A and B of the universal set U,

① $n(A^C) = n(U) - n(A)$

② $n(A - B) = n(A) - n(A \cap B) = n(A \cup B) - n(B)$

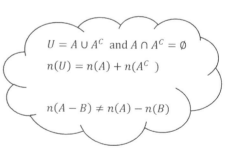

Note $R = \{x \mid x \text{ is a real number}\}$

$Q = \{x \mid x \text{ is a rational number}\}$

$Z = \{x \mid x \text{ is an integer}\}$

$N = \{x \mid x \text{ is a natural number}\}$

$\Rightarrow \; N \subset Z \subset Q \subset R$

Exercises

#1 Identify all the sets. Mark o for a set or × for a non-set.

    (1) A set of pretty girls in a school.

    (2) A set of red apples.

    (3) A set of natural numbers.

    (4) A set of small numbers.

    (5) A set of famous singers.

    (6) A set of even numbers.

    (7) A set of people who like math.

    (8) A set of students whose heights are less than 5 feet in a class.

    (9) The set of 1-digit odd numbers.

    (10) The set of healthy foods in a store.

#2 Determine whether the following notations are true or false.

    (1) $\{1, 2, 3\} = \{3, 1, 2\}$

    (2) $\{1, 2, 3, 4, 5\} = \{x \mid x \text{ is a natural number less than 6.}\}$

    (3) $\{x \mid x \text{ is a factor of 6.}\} = \{1, 2, 3, 6\}$

    (4) $\{x \mid x \text{ is a natural number less than 1.}\} = \{0\}$

    (5) $\{0, 4, 8, 12, 16 \cdots\} = \{x \mid x \text{ is a multiple of 4.}\}$

    (6) $\{1, 3, 5, 7, 9\} = \{x \mid x \text{ is an odd number less than 10.}\}$

    (7) $\{x \mid 1 \leq x \leq 3, \ x \text{ is an integer.}\} = \{1, 2, 3\}$

    (8) $\{1, 2, 3, 4\} = \{x \mid x \text{ is a prime number less than 5.}\}$

#3 State if the following sets are finite or infinite sets.

    (1) $\{x \mid x \text{ is a factor of 20.}\}$

    (2) $\{x \mid x \text{ is a multiple of 2.}\}$

    (3) $\{x \mid x \text{ is an even number.}\}$

    (4) $\{x \mid 1 \leq x \leq 3, \ x \text{ is an odd number.}\}$

    (5) $\{x \mid x^2 + 1 = 0, \ x \text{ is a real number.}\}$

    (6) $\{x \mid x \text{ is an odd number bigger than 10.}\}$

#4 Find the value of $n(A) + n(B)$ for the following sets A and B

(1) $A = \{x \mid x$ is a factor of 10. $\}$, $B = \{x \mid x$ is an even number less than 10. $\}$

(2) $A = \{x \mid 1 \leq x \leq 5,\ x$ is an odd number. $\}$, $B = \{0\}$

(3) $A = \{x \mid x$ is a natural number less than 1. $\}$, $B = \{x \mid 3 < x < 4,\ x$ is a natural number. $\}$

(4) $A = \{1, 2, 3, 4\}$, $B = \{2a + 1 \mid a \in A\}$

#5 Find the value of $a + b$ for the following sets A and B

(1) $A = \{1, 2, a + 3\}$ and $B = \{2, 5, b + 1\}$, $A \subset B$ and $B \subset A$

(2) $A = \{3, 4, a + 1\}$ and $B = \{a + 2, b, 4\}$, $A \subset B$ and $B \subset A$

#6 Find the number of subsets and proper subsets for the following sets A

(1) $A = \{x \mid x$ is a factor of 15. $\}$

(2) $A = \{x \mid x$ is an even number less than 8. $\}$

#7 $A = \{x \mid x$ is a factor of 12. $\}$ and $B = \{x \mid x$ is a factor of 6. $\}$

How many number of subsets which include all the elements of B are in the subsets of A?

#8 How many number of subsets which include the element a but not the elements b and c are in the subsets of $A = \{a, b, c, d, e, f\}$?

#9 Find the number of a set A which satisfies the conditions for (1), (3), and (4).

For (2), find the set A.

(1) $\{1\} \subset A \subset \{1, 2, 3\}$

(2) $\{2, 3\} \subset A \subset \{2, 3, 4, 5, 6\}$ and $n(A) = 3$

(3) $A \subset \{x \mid x$ is a natural number less than 5. $\}$ and A has at least one even number.

(4) $A \subset \{x \mid x$ is a factor of 20. $\}$ and $(1 \in A$ or $2 \in A$)

#10 Find the value of $p + q$.

(1) The number of subsets of A is 64 and $n(A) = p$.

The number of proper subsets of B is 7 and $n(B) = q$.

(2) $A \subset \{x \mid 1 \leq x \leq p + q, \ x$ is a natural number. $\}$,

$(p \in A$ and $q \in A$), and $n(A) = 32$, where $p + q > 2$

(3) $A \subset \{x \mid 1 \leq x \leq p + q, \ x$ is a natural number. $\}$,

$(p \in A$ and $p + q \in A)$, $1 \notin A$, and $n(A) = 32$, where $p + q > 3$

#11 Find the intersection of the following sets

(1) $A = \{x \mid x$ is a factor of 6. $\}$, $B = \{x \mid 1 \leq x \leq 10, \ x$ is an even number. $\}$

(2) $A = \{1, 2, 3, 4, 5\}$, $B = \{x \mid x = a + 1, \ a \in A \}$

(3) $A = \{x \mid x$ is a multiple of 3. $\}$, $B = \{x \mid x$ is a factor of 12. $\}$

(4) $A = \{x \mid x$ is an even number. $\}$, $B = \{x \mid x$ is an odd number. $\}$

#12 Find the set A which satisfies the following conditions

(1) $B = \{1, 2, 3\}$, $A \cup B = \{1, 2, 3, 4, 5\}$, $A \cap B = \emptyset$

(2) $B = \{1, 2, 3, 4\}$, $A \cup B = \{1, 2, 3, 4, 5, 6\}$, $A \cap B = \{1, 2\}$

(3) $B = \{1, 2, 3\}$, $A \cup B = \{1, 2, 3, 4, 5\}$, $n(A \cap B) = 2$

(4) $A = \{a, a + 1, a + 2\}$, $B = \{3, 4, 5, 6, 7\}$, $A \cap B = \{3, 4\}$

#13 Find the number of a set A with the following conditions

(1) $B = \{x \mid x$ is a factor of 6. $\}$, $C = \{x \mid x$ is a factor of 18. $\}$, $A \cap B = B$, $A \cup C = C$

(2) For a set B, $n(B) = 5$, $n(A \cap B) = 3$, and $n(A \cup B) = 10$

(3) For a set B, $A \cap B = \emptyset$, $n(B) = 7$, and $n(A \cup B) = 15$

(4) For a set B, $n(A \cup B) = 20$, $n(A \cap B) = 5$, and $n(B - A) = 8$

(5) For a fixed set $U = \{a, b, c, d, e, f, g\}$ and a set B,

$A - B = \{a, b\}$, $B - A = \{c, d\}$, and $(A \cup B)^C = \{f\}$

(6) For two sets B and C, $n(B \cup C) = 10$, $n(B \cap C) = 3$, $n(C) = 5$, and $A \subset B$, $A \cap C = \emptyset$

#14 Find the sets $A - B$ and $B - A$ for the following sets A and B.

(1) $A = \{1, 2, 3, 4, 5\}$, $B = \{x \mid x \text{ is a factor of 4.}\}$

(2) $A = \{1, 2, 3, 4, 5, 6\}$, $B = \{x \mid 1 \leq x \leq 7, \ x \text{ is an odd number.}\}$

(3) For a fixed set $U = \{x \mid x \text{ is a factor of 20.}\}$,

$A \subset U, \ B \subset U, \ A \cap U = \{2, 5, 10\}, \ A \cap B = \{5, 10\}, \ (A \cup B)^C = \{1\}$

#15 Solve the following operations for any subsets A and B of the fixed set U.

(1) A^C

(2) $(A^C)^C$

(3) $U - A^C$

(4) $A - A^C$

(5) $(A \cup A^C) - (A \cap A^C)$

(6) $A \cap B^C$

(7) $A^C \cap B^C$

(8) $A - B$ when $A \cap B = A$

(9) $A - B$ when $A \cap B = \emptyset$

(10) $A \cap B$ when $A - B = A$

(11) $A^C - B^C$ when $A \subset B$

Chapter 19. Probability

19-1 Probability

1. Measuring Probability

> If all of the outcomes have the same chance of occurrence, they are called *equally likely*.
> Example : When we toss a coin or throw a dice, we shall assume the possible outcomes are equally likely if not others mentioned.

(1) Probability (Classical Definition)

For some problems, there are no clear answers.

Probability deals with such randomness and uncertainty in mathematics.

Think about tossing a single coin.

If somebody asks, "What is the probability of obtaining heads? ", we usually answer that it is 50%, $\frac{1}{2}$, 0.5, or 1:2.

The probability is the ratio of the number of times a head obtains to the total number of outcomes (head, tail) in an experiment.

Suppose none of the outcomes (various results) occur at the same time.

If an experiment is repeated equally likely large number (N) of times, then the *probability* of an outcome (an even E), symbolized $P(E)$ is represented by

$$P(E) = \frac{N_E}{N} \quad \text{, where } N_E \text{ is the number of times an event } E \text{ occurs.}$$

> $P(E) = \dfrac{\text{the number of times an event } E \text{ occurs}}{\text{the total number of possible outcomes}}$

(2) Probability (Relative Frequency)

Tossing a single coin, the relative frequency of heads is quite changeable in two, three, or ten tosses of a coin. But after tossing a coin thousand times, it remains stable. If in many tosses of a coin, the proportion of heads observed is close to $\frac{1}{2}$, then we say that the probability of a head on a toss is $\frac{1}{2}$. Thus, we say probability is long-term relative frequency.

If an experiment is performed large number (N) of times with outcomes (N_E) and the relative frequency $\frac{N_E}{N}$ approaches a limiting numerical number as the number N increases, then the probability of an outcome, $P(E)$, is represented by

$$P(E) = \lim_{n \to \infty} \frac{N_E}{N} \quad \text{, where } N_E \text{ is the number of times an event } E \text{ occurs.}$$

Using the relative frequency, we can estimate the probability of particular outcome.

The estimate of probability will be more accurate if more experiments are performed.

2. Sample Spaces and Events

(1) Sample Spaces and Sample Points

The *sample space* of an event is a set of all logical possible outcomes for a probability experiment and is denoted by $S = \{E_1, E_2, E_3, \cdots\cdots, E_n\}$.

Each element, $E_1, E_2, E_3, \cdots\cdots, E_n$ in a sample space is called a *sample point*.

Example

In a single flip of a coin, the sample space is $\{H, T\}$.

The sample space of flipping two coins is $S = \{(H, H), (H, T), (T, H), (T, T)\}$

Where $E_1 = \{(H, H)\}$: both coins are heads,

$E_2 = \{(H, T)\}$: the first coin is a head and the second is a tail,

$E_3 = \{(T, H)\}$: the first coin is a tail and the second is a head,

$E_4 = \{(T, T)\}$: both coins are tails.

(2) Events

Any collection (subset) of outcomes contained in the sample space for a probability experiment is called an *event*.

1) Simple Event

A *simple event* consists of exactly one outcome.

In this case, we usually write $E_1, E_2, E_3 \cdots\cdots$ instead of $\{E_1\}, \{E_2\}, \{E_3\}, \cdots\cdots$

2) Compound Event

A *compound event* consists of more than one event.

Example

For the sample space $S = \{(H, H), (H, T), (T, H), (T, T)\}$ of flipping two coins, there are 4 simple events, $E_1 = \{(H, H)\}$, $E_2 = \{(H, T)\}$, $E_3 = \{(T, H)\}$, $E_4 = \{(T, T)\}$, and some compound events including

$E = \{(H, H), (H, T), (T, H)\}$ (the event that at least one head shows) and $F = \{(H, T), (T, H), (T, T)\}$ (the event that at least one tail shows).

3. The Relationship between an Event and a Set

For any two events E and F of a sample space S,

(1) Union

The *union* of two events E and F, denoted by $E \cup F$, is the event consisting of all outcomes that are either in E or in F or in both E and F. The event $E \cup F$ will occur if either E or F occurs.

(2) Intersection

The *intersection* of two events E and F, denoted by $E \cap F$, is the event consisting of all outcomes that are in both E and F. The event $E \cap F$ will occur only if both E and F occur.

(3) Complement

The *complement* of an event E, denoted by E^C, is the set of all outcomes in S that are not included in E.

Example

Consider rolling a die.

Let $A = \{2, 4, 6\}$, $B = \{1, 3, 5\}$, $C = \{1, 2, 3, 4\}$. Then,

$A \cup B = \{1, 2, 3, 4, 5, 6\}$, $A \cap B = \emptyset$, $A \cap C = \{2, 4\}$, $A^C = \{1, 3, 5\}$, $(A \cup C)^C = \{5\}$.

(4) Mutually Exclusive (Disjoint) Events

Two events E and F are mutually exclusive if and only if $E \cap F = \emptyset$, where the null event, denoted by $\emptyset$, consists of no points.

That is, if two events E and F have no common outcomes, then E and F are *mutually exclusive* or *disjoint events*.

Example

If $E = \{(H, H), (T, T)\}$ and $F = \{(H, T), (T, H)\}$, then $E \cap F = \emptyset$.

So, the two events E and F are mutually exclusive.

Note: Venn Diagrams

  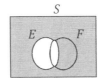

Shaded region is $E \cup F$.   Shaded region is $E \cap F$.   Shaded region is E^C. E and F are mutually exclusive events.

4. Properties of Probability

Every random experiment has a sample space $S = \{E_1, E_2, E_3, \cdots\cdots, E_n\}$.

To each simple event E_i, $i = 1, 2, \cdots\cdots, n$, a probability $P(E_i)$, $i = 1, 2, \cdots\cdots, n$ is assigned.

---

(1) For each simple event E_i, $i = 1, 2, \cdots\cdots, n$, $0 \leq P(E_i) \leq 1$, $i = 1, 2, \cdots\cdots, n$ and $P(S) = 1$.

If $E_1, E_2, E_3, \cdots\cdots, E_n$ is a finite collection of mutually exclusive events, then

$$P(E_1 \cup E_2 \cup E_3 \cup \cdots\cdots \cup E_n) = P(E_1) + P(E_2) + P(E_3) + \cdots\cdots + P(E_n).$$

If $E \cap F = \emptyset$, then $P(E \cup F) = P(E) + P(F)$

---

(2) For any event E, $P(E) = 1 - P(E^C)$

($\because$ If E and E^C are mutually exclusive events, then $E \cup E^C = S$.

Since $P(S) = 1$, $1 = P(S) = P(E \cup E^C) = P(E) + P(E^C)$.

Therefore, $P(E) = 1 - P(E^C)$.)

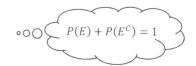

$P(E) + P(E^C) = 1$

---

(3) For any events E and F, if $E \cap F = \emptyset$, then $P(E \cap F) = 0$

($\because$ If an event $E \cap F$ contains no outcomes, then $(E \cap F)^C = S$.

Since $P(S) = 1$, $1 = P(S) = P((E \cap F)^C) = 1 - P(E \cap F)$.

Therefore, $P(E \cap F) = 0$.)

---

(4) For any events E and F, $P(E \cup F) = P(E) + P(F) - P(E \cap F)$

($\because$ Note that $E \cup F = E \cup (F \cap E^C)$ in a Venn Diagram.

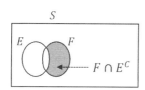

Since $E \cap (F \cap E^C) = \emptyset$, $P(E \cup F) = P(E) + P(F \cap E^C)$.

Note that $F = (E \cap F) \cup (F \cap E^C)$ in a Venn Diagram.

Since $(E \cap F) \cap (F \cap E^C) = \emptyset$, $P(F) = P(E \cap F) + P(F \cap E^C)$.

$\therefore P(F \cap E^C) = P(F) - P(E \cap F)$.

Therefore, $P(E \cup F) = P(E) + P(F) - P(E \cap F)$.)

Example For a selected card from a standard deck of 52 cards, find the probability that the card is either a heart or a face card.

Note that the deck has 13 hearts and 12 face cards.

Let A be the event of selecting a heart and B be the event of selecting a face card.

Then, $P(A) = \dfrac{13}{52}$ and $P(B) = \dfrac{12}{52}$.

Since three of the cards are hearts and faces, $P(A \cap B) = \dfrac{3}{52}$

Therefore, $P(A \cup B) = P(A) + P(B) - P(A \cap B) = \dfrac{13}{52} + \dfrac{12}{52} - \dfrac{3}{52} = \dfrac{22}{52}$

5. Counting Outcomes

(1) Tree Diagrams

A tree diagram is used to display all possible outcomes of experiments with more than one event.

Example

The sample space of flipping two coins is $S = \{(H, H), (H, T), (T, H), (T, T)\}$.

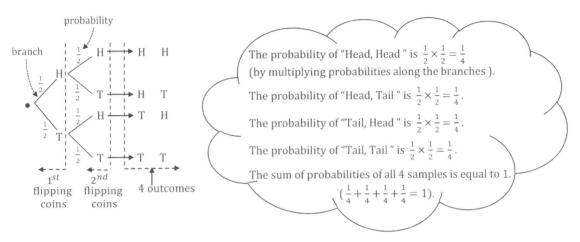

There are 2 branches for the 1st set of flipping coins and for each branch, there are 2 branches for the 2nd set of flipping coins.

So, the total number of outcomes of the two experiments is $2 \times 2 = 4$.

(2) The Multiplication Principle

1) The Basic Rule of Counting

Given two experiments, if Experiment 1 has m possible outcomes, and if, for each outcome of Experiment 1, the Experiment 2 has n possible outcomes, there is a total of $m \times n$ possible outcomes of the two experiments.

2) The Generalized Basic Rule of Counting

Given n experiments, if the first experiment has m_1 possible outcomes, and if for each m_1 outcome the second experiment has m_2 possible outcomes, and if for each m_1 and m_2 outcomes the third experiment has m_3 possible outcomes, and if $\cdots\cdots$,

then there is a total of $m_1 \times m_2 \times m_3 \cdots\cdots \times m_n$ possible outcomes of the n experiments.

Example

When flipping three coins,

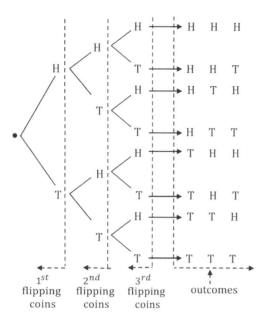

The total possible outcome of the 3 experiments is $2 \times 2 \times 2 = 2^3 = 8$.

Note:

Number of flipping coins	Total possible outcomes
1	2^1
2	2^2
3	2^3
$\vdots$	$\vdots$
m	2^m

When rolling n dice $\Rightarrow$ 6^n *outcomes*

When flipping m coins and rolling n dice $\Rightarrow$ $2^m \times 6^n$ *outcomes*

6. Permutations

Suppose we want to find the number of ways to arrange objects in which order does matter.

For letters $A, B,$ and $C,$ there are 6 different ordered arrangements, $ABC, \ ACB, \ BAC, \ BCA,$ $CAB, \ CBA.$

Each arrangement is called a *permutation* of the three letters $A, B,$ and $C.$

The number of possible permutation can be identified using the basic rule of counting.

We place the three possible letters for the first object. Each of these choices leaves two remaining choices for the second object, and only one choice remaining for the third object. This means there are $3 \cdot 2 \cdot 1 = 6$ possible different orders (permutations).

Now, consider n (a positive integer) objects.

In this case, there are $(n) \cdot (n-1) \cdot (n-2) \cdots (3) \cdot (2) \cdot (1)$ possible different permutations.

If a positive integer n is multiplied by all the preceding positive integers, the result is called *n factorial* and is denoted by $n!$ ($n! = nPn$).

That is,

$$n \cdot (n-1) \cdot (n-2) \cdots 3 \cdot 2 \cdot 1 = n!$$

: The product of all positive integers less than or equal to n

Example $1! = 1$

$2! = 2 \cdot 1 = 2$

$3! = 3 \cdot 2 \cdot 1 = 6$

$4! = 4 \cdot 3 \cdot 2 \cdot 1 = 24$

> $0!$ is defined to be 1.
> That is, $0! = 1$

> $\dfrac{10!}{7!} = \dfrac{10 \cdot 9 \cdot 8 \cdots 2 \cdot 1}{7 \cdot 6 \cdot 5 \cdots 2 \cdot 1}$
> $= 10 \cdot 9 \cdot 8 = 720$

The number of different ways of r objects that can be taken from a set of n obhects is symbolized by the notation nPr (the permutation of r objects taken from n objects) and is defined as the product of all positive integers r greatest factors of $n!$

That is, $\boxed{nPr = \dfrac{n!}{(n-r)!} = \dfrac{n \cdot (n-1) \cdot (n-2) \cdots 3 \cdot 2 \cdot 1}{(n-r) \cdot (n-r-1) \cdot (n-r-2) \cdots 3 \cdot 2 \cdot 1} = n \cdot (n-1) \cdot (n-2) \cdots (n-r+1)}$

Example   How many different ways of 3 letters could be selected from a total of 10 letters without replacement?

$10P3 = 10 \cdot 9 \cdot 8 = 720$ different ways

We can use permutations if all objects are chosen from the same group, if no object may be used more than once, and if the order of arrangement matters.

For example, consider the different letter arrangements formed by the letters MIRROR.

In this case, there are

$$\frac{6!}{1! \cdot 1! \cdot 3! \cdot 1!} = \frac{6 \cdot 5 \cdot 4 \cdot 3!}{1! \cdot 1! \cdot 3! \cdot 1!} = 6 \cdot 5 \cdot 4 = 120 \text{ possible letter arrangements.}$$

In the case of PEPPER, there are

$\frac{6!}{3!\cdot 2!} = \frac{6\cdot 5\cdot 4\cdot 3!}{3!\cdot 2} = 60$ possible letter arrangements.

If in a set of n objects, r objects are identical, there are $\frac{n!}{r!}$ different ways.

Example How many different ways, each consisting of 7 flags hung in a line, can be arranged from a set of 3 red flags, 2 white flags, and 2 blue flags if all flags of the same color are identical?

$\frac{7!}{3!2!2!} = \frac{7\cdot 6\cdot 5\cdot 4\cdot 3\cdot 2\cdot 1}{(3\cdot 2\cdot 1)(2\cdot 1)(2\cdot 1)} = 7\cdot 5\cdot 3\cdot 2\cdot 1 = 210$ different ways

7. Combinations

$nCr = \frac{n!}{(n-r)!\,r!} = \frac{n(n-1)(n-2)\cdots\cdots(n-r+1)}{r!}$

The order of the elements is important in permutations.

In contrast, a combination is a selection of r objects taken without regard to the order of the objects.

The number of different ways of r objects that can be taken from a set of n objects is symbolized by the notation nCr (the combination of r objects taken from n objects) for $r \leq n$.

We define

$$nCr = \binom{n}{r} = \frac{nPr}{r!} = \frac{n!}{(n-r)!\,r!} \quad \text{for } r \leq n$$

: The number of possible combinations of n objects taken r at a time.

Example A group of 5 is to be formed from a class of 20 students.

How many different groups are possible?

There are $20C5 = \frac{20P5}{5!} = \frac{20!}{(20-5)!\,5!} = \frac{20\cdot 19\cdot 18\cdot 17\cdot 16\cdot 15!}{15!(5\cdot 4\cdot 3\cdot 2\cdot 1)} = 19\cdot 3\cdot 17\cdot 16 = 15,504$ groups.

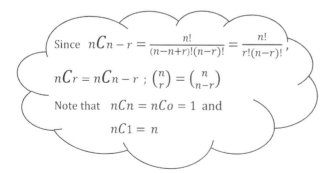

Since $nCn-r = \frac{n!}{(n-n+r)!(n-r)!} = \frac{n!}{r!(n-r)!}$,

$nCr = nCn-r$; $\binom{n}{r} = \binom{n}{n-r}$

Note that $nCn = nC0 = 1$ and

$nC1 = n$

If the order does not matter, it is a combination.

If the order does matter, it is a permuattion.

19-2 Conditional Probability and Independence/Dependence

1. Conditional Probability

For any two events E and F with $P(F) > 0$,

the conditional probability of E given that F has occurred is defined by

$$P(E \setminus F) = \frac{P(E \cap F)}{P(F)}$$

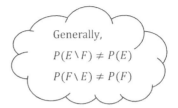

Generally,
$P(E \setminus F) \neq P(E)$
$P(F \setminus E) \neq P(F)$

Note: $\boxed{P(E \cap F) = P(E \setminus F) \cdot P(F)}$

2. Independence and Dependence

If the knowledge that F has occurred does not change the probability that E occurs or has

occurred, then $P(E \setminus F) = P(E)$. In this case, we say that E is *independent* of F.

Since $P(E \setminus F) = \frac{P(E \cap F)}{P(F)}$,

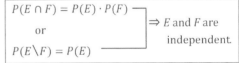

$P(E \cap F) = P(E) \cdot P(F)$
or
$P(E \setminus F) = P(E)$
$\Rightarrow$ E and F are independent.

$\boxed{E \text{ is independent of } F \text{ if } P(E \cap F) = P(E) \cdot P(F)}$

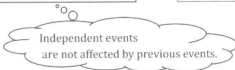

Independent events
are not affected by previous events.

If two events E and F are not independent: that is, the outcome of one event does affect the

outcome of the other, then we say that E and F are *dependent*.

Note: ① *To find the probability of two independent events both occurring,*

   multiply the probability of the first event (E) by the probability of the second event (F).

   $P(E \text{ and } F) = P(E) \cdot P(F)$

② *To find the probability of two dependent events both occurring,*

   multiply the probability of the first event (E) by the probability of the second event (F)

   after E occurs.

   $P(E \text{ and } F) = P(E) \cdot \underline{P(F \text{ following } E)}$

   ⌐------ *the conditional probability of F given that E has occurred*

Example

Consider flipping a coin and rolling a die.

The probability of obtaining heads in one coin flip is $\frac{1}{2}$ and the probability of obtaining a number 3 in one die roll is $\frac{1}{6}$. In this case, the occurrence of one event does not affect the occurrence of the other. So, the two events are independent.

Therefore, the probability of both occurring is $\frac{1}{2} \cdot \frac{1}{6} = \frac{1}{12}$.

Example

Consider rolling a die.

Define the events $E = \{1, 2, 3\}$, $F = \{2, 4, 6\}$, and $G = \{1, 3, 5, 6\}$.

Then, $P(E) = \frac{3}{6} = \frac{1}{2}$, $P(E \backslash F) = \frac{P(E \cap F)}{P(F)} = \frac{\frac{1}{6}}{\frac{3}{6}} = \frac{1}{3}$, and $P(E \backslash G) = \frac{P(E \cap G)}{P(G)} = \frac{\frac{2}{6}}{\frac{4}{6}} = \frac{1}{2}$.

Since $P(E) = P(E \backslash G) = \frac{1}{2}$, events E and G are independent.

Since $P(E) \neq P(E \backslash F)$, events E and F are dependent.

Exercises

#1 Find the sample space for the following

    (1) A spin of a spinner marked 0, 1, 2, 3

    (2) A toss of one coin and a spin of a spinner marked 0, 1, 2, 3

    (3) Two coins tossed once

    (4) A pair of dice tossed once

    (5) A toss of one coin and a toss of an ordinary die.

#2 Think about tossing one coin and one ordinary die. Find the probability of the event

    (1) E : Two heads occur.

    (2) F : A head and an even number occur.

    (3) G : An odd or even number occurs.

#3 Find the probabilities for the indicated sample spaces derived from the random experiment of drawing one card from a full deck.

    (1) $S = \{\text{red}, \text{ black}\}$

    (2) $S = \{\text{ace or picture card}, \text{ otherwise}\}$

#4 One ball is drawn from a bag containing three red balls marked 1, 2, 3 ; four blue balls marked 1, 2, 3, 4 ; and two yellow balls marked 1, 2. Find the probability for the indicated sample spaces from this random experiment.

    (1) $S = \{\text{red}, \text{ blue}, \text{ yellow}\}$

    (2) $S = \{\text{even}, \text{odd}\}$

#5 Three coins are tossed at once. Find $P(E \cap F)$.

    (1) Let E be the event "coins match" and let F be the event "not more than one head".

    (2) Let E be the event "coins match" and let F be the event "not more than three heads".

    (3) Let E be the event "coins match" and let F be the event "at least two heads".

(4) Let E be the event "coins match" and let F be the event "at least one head".

(5) Let E be the event "head on first toss" and let F be the event "tail on second toss".

#6 For the following experiments, one toss is made. Find the probabilities indicated.

(1) Two coins, E: "at most one head", F : "no tails". Find $P(E \cup F)$

(2) Three coins, E: "at least two heads", F : "only one tail". Find $P(E \cup F)$.

(3) Two dice, $E = \{(1,2)\}$, $F = \{(3,4)\}$. Find $P(E \cup F)$.

(4) Two dice, E: "The sum of marked numbers is 6". Find $P(E)$.

(5) Two dice, E: "sum $\leq$ 10". Find $P(E)$.

#7 Find the number of different arrangements (permutations) for the following events

(1) Scheduling seven different classes in seven periods.

(2) Creating the batting order for a baseball team consisting of 9 players.

#8 A coin is flipped twice. What is the conditional probability that both coins are heads, given that the first coin is a head?

#9 A bag contains 5 red, 7 white, and 10 black balls. A ball is chosen at random from the bag, and it is noted that it is not one of the white balls. What is the conditional probability that it is red?

#10 A box contains 10 apples and 15 pears. The fruits to be chosen are selected at random. Find the probability that

(1) The first two fruits chosen are apples.

(2) The second fruit chosen is an apple.

(3) Given that the second fruit chosen is an apple, the first fruit chosen is also an apple.

#11   Think about tossing a coin and an ordinary dice. Let E be the event " H on coin " and let F be the event " 2 on dice ".   Find $P(E \cup F), P(E \cap F),$ and determine if E and F are dependent or independent.

#12   Think about tossing a nickel and a dime.   Let E be the event " coins match ", let F be the event " nickel falls on heads " , and let G be the event " at least one head shows ".
Determine which events are independent.

Chapter 1. The Natural Numbers

#1 (1) 1, 2, 4 / 0, 4, 8, 12, 16, ⋯   (2) 1, 7 / 0, 7, 14, 21, ⋯   (3) 1, 2, 3, 4, 6, 12 / 0, 12, 24, 36, ⋯

(4) 1, 2, 3, 4, 6, 9, 12, 18, 36 / 0, 36, 72, 108, ⋯   (5) 1 / 0, 1, 2, 3, 4, 5, ⋯

(6) 1, 3, 5, 15 / 0, 15, 30, 45, ⋯   (7) 1, 2, 3, 6, 9, 18 / 0, 18, 36, 54, ⋯

(8) 1, 2, 3, 4, ⋯ / undefined

#2 (1) False   (2) False   (3) True   (4) False   (5) False   (6) True   (7) False   (8) True

(9) True   (10) False   (11) False   (12) False

#3 (1) 9   (2) 6   (3) 3   (4) 8

#4 (1) 2, 3   (2) 3, 7   (3) 2, 7   (4) 2, 3, 5   (5) 2, 3, 5, 7   (6) 2, 5

#5 (1) 6   (2) 15   (3) 42   (4) 70   (5) 6   (6) 32   (7) 11   (8) 18

#6 (1) 4   (2) 8   (3) 9   (4) 12   (5) 9

#7 (1) 2   (2) 2   (3) 4

#8 (1) GCF = $2^2 \cdot 3 \cdot 5$ LCM = $2^3 \cdot 3^2 \cdot 5 \cdot 7$ (2) GCF = $2^2 \cdot 3$ LCM = $2^4 \cdot 3^3 \cdot 5^2 \cdot 7 \cdot 11$

(3) GCF = $2 \cdot 3^2$ LCM = $2^3 \cdot 3^4 \cdot 5^2 \cdot 7$ (4) GCF = 3   LCM = $2^2 \cdot 3^3 \cdot 5 \cdot 7^2$

#9 (1) 6   (2) 6

#10 (1) 1, 2, 3, 4, 6, 12   (2) 1, 2, 3, 6   (3) 24, 48, 72, ⋯   (4) 45, 90, 135, ⋯

(5) 18, 36, 54, ⋯ or $\frac{18}{5}, \frac{36}{5}, \frac{54}{5}, \cdots$

#11 (1) 36   (2) 120   (3) 1260   (4) 56

Chapter 2. Integers and Rational Numbers

#1 (1) $+30\%$ (2) -5 points (3) $+1$ week or $+7$ days (4) -3 degree

#2

$$\xleftarrow{\quad}\overset{\displaystyle\underset{\displaystyle C}{\bullet}}{\underset{-5}{\quad}}\ \underset{-4}{|}\ \underset{-3}{|}\ \overset{\displaystyle\underset{\displaystyle B}{\bullet}}{\underset{-2}{\quad}}\ \overset{\displaystyle\underset{\displaystyle F}{\bullet}}{\underset{-1}{\quad}}\ \overset{\displaystyle\underset{\displaystyle E}{\bullet}}{\underset{0}{\quad}}\ \underset{1}{|}\ \overset{\displaystyle\underset{\displaystyle A}{\bullet}}{\underset{2}{\quad}}\ \overset{\displaystyle\underset{\displaystyle D}{\bullet}}{\underset{3}{\quad}}\xrightarrow{\quad}$$

#3 $-8,\ -5,\ 0,\ 1,\ 4,\ 5$

#4 (1) 2 (2) 3 (3) 3 (4) 1

#5 (1) -1 (2) -7 (3) -6 (4) 12 (5) 5 (6) 1 (7) $-\dfrac{2}{3}$ (8) $\dfrac{3}{4}$ (9) -5 (10) 9

   (11) 24 (12) 6 (13) 6

#6 (1) 1 (2) -5 (3) 0 (4) 8 (5) 0 (6) -4 (7) -2 (8) -3 (9) -12 (10) 3 (11) 0

#7 (1) Positive (2) Negative (3) Positive (4) Positive (5) Negative (6) Positive

   (7) Positive (8) Negative

#8 (1) 4 (2) 8 (3) 14

#9 (1) -7 (2) -70 (3) 60 (4) -1 (5) 0

#10 (1) False (2) False (3) True (4) True (5) True (6) False (7) False (8) True

   (9) False (10) True

#11 (1) $-7 < -\dfrac{1}{4} < 0 < \dfrac{1}{3} < \dfrac{1}{2}$ (2) $-5 < -2 < \dfrac{1}{4} < \dfrac{3}{5} < 2$ (3) $-\dfrac{1}{2} < -\dfrac{1}{5} < \dfrac{7}{12} < \dfrac{3}{4} < \dfrac{5}{6}$

   (4) $-\dfrac{3}{4} < -\dfrac{2}{3} < -\dfrac{1}{6} < \dfrac{5}{8} < \dfrac{4}{6}$

#12 (1) $\dfrac{3}{2} \neq \dfrac{5}{8}$ (2) $\dfrac{3}{4} \neq \dfrac{4}{7}$ (3) $\dfrac{0}{3} = \dfrac{0}{5}$ (4) $\dfrac{6}{5} \neq \dfrac{3}{2}$ (5) $\dfrac{2}{3} = \dfrac{10}{15}$ (6) $\dfrac{4}{9} = \dfrac{16}{36}$ (7) $\dfrac{3}{7} \neq \dfrac{14}{28}$

   (8) $3 = \dfrac{12}{4}$ (9) $\dfrac{5}{24} \neq \dfrac{4}{16}$ (10) $\dfrac{5}{12} \neq \dfrac{7}{16}$

#13 (1) $\dfrac{23}{6}$ (2) $-\dfrac{22}{5}$ (3) $\dfrac{43}{8}$ (4) $-\dfrac{20}{3}$ (5) $\dfrac{37}{5}$ (6) $-\dfrac{8}{3}$ (7) $\dfrac{13}{3}$ (8) $\dfrac{25}{7}$ (9) $-\dfrac{21}{4}$ (10) $\dfrac{14}{5}$

#14 4 is the greatest number of $a - b$ and -9 is the smallest number of $a + b$.

#15 $-\dfrac{13}{20},\ -\dfrac{11}{20},\ -\dfrac{9}{20},\ -\dfrac{7}{20},\ -\dfrac{3}{20},\ -\dfrac{1}{20},\ \dfrac{1}{20},\ \dfrac{3}{20},\ \dfrac{7}{20},\ \dfrac{9}{20},\ \dfrac{11}{20}$

#16 (1) $5\frac{13}{20}$ (2) $4\frac{5}{12}$ (3) $\frac{27}{28}$ (4) $8\frac{1}{24}$ (5) $1\frac{38}{45}$ (6) $6\frac{1}{10}$ (7) $-\frac{1}{3}$ (8) $\frac{11}{12}$

#17 (1) $8\frac{2}{5}$ (2) $6\frac{4}{7}$ (3) $\frac{11}{30}$ (4) $3\frac{1}{3}$ (5) $10\frac{2}{5}$ (6) $\frac{8}{15}$ (7) $1\frac{7}{15}$ (8) $10\frac{1}{2}$ (9) $2\frac{11}{12}$ (10) $1\frac{5}{6}$

(11) $\frac{2}{3}$ (12) $1\frac{1}{5}$ (13) $\frac{3}{4}$

#18 (1) $-\frac{3}{4}$ (2) $-1\frac{7}{20}$ (3) $-2\frac{1}{12}$ (4) 7 (5) 49 (6) $-3\frac{23}{24}$ (7) $8\frac{5}{6}$ (8) $-3\frac{29}{36}$ (9) $-\frac{1}{24}$

(10) $\frac{23}{24}$ (11) $\frac{17}{44}$ (12) $-2\frac{1}{6}$ (13) 14 (14) $7\frac{1}{3}$ (15) $\frac{2}{3}$ (16) 25 (17) 7 (18) $16\frac{3}{4}$

(19) $6\frac{1}{4}$ (20) $1\frac{7}{52}$

#19 $\left(-\frac{1}{a}\right)^2 > -\frac{1}{a} > -a > a^2 > \frac{1}{a} > -\frac{1}{a^2}$

#20 (1) 1 (2) $\frac{1}{4}$ (3) $-\frac{5}{12}$ (4) 77 (5) $\frac{4}{7}$

Chapter 3. Equations

#1 (1) 0 (2) 4 (3) -4 (4) -16 (5) 19 (6) -26 (7) $-\dfrac{5}{6}$ (8) -2 (9) -6 (10) $\dfrac{1}{2}$

(11) -5 (12) -2

#2 (1) $-3x$ (2) $-4x+6$ (3) $-20x+\dfrac{20}{3}$ (4) $-4x-2\dfrac{1}{2}$ (5) $x-5$ (6) $-x+2y+2$

(7) $-2a-b-4$ (8) $4m^2-3m+1$ (9) $\dfrac{3}{2}t^2-2t$ (10) $2a-3b-5$ (11) $-4x+10$

(12) $\dfrac{1}{3}x-1\dfrac{11}{12}$

#3 $4(a+b)$ **#4** $3x-7$ **#5** (4)   **#6** $a+b=\dfrac{1}{6}$ and $a-b=\dfrac{5}{2}$

#7 $a\cdot b=-8$ and $\dfrac{a}{b}=-\dfrac{1}{8}$ **#8** $9x+5$

#9 (1) 3 (2) 5 (3) $\dfrac{1}{6}$ (4) 1 (5) $-\dfrac{7}{25}$ (6) -2 (7) 3 (8) $\dfrac{6}{11}$

#10 3 **#11** $\dfrac{a-c}{b}$ **#12** -1 **#13** $1\dfrac{7}{12}$ **#14** 3 **#15** -6 **#16** 2

#17 0 **#18** 6 **#19** $\dfrac{7}{8}$ **#20** $-\dfrac{2}{5}$ **#21** $-3\dfrac{2}{5}$ **#22** $-\dfrac{1}{36}$

#23 10 ounces of water   **#24** $\dfrac{5}{4}$ ounces of salt   **#25** 50 minutes   **#26** $31\dfrac{1}{4}$ ounces

#27 12 dollars   **#28** $\$66.59$ **#29** 12 years old   **#30** 53

#31 37   **#32** 2 years later   **#33** $1\dfrac{1}{5}$ hours   **#34** 2 days   **#35** 180 pages

#36 2 hours 30 minutes   **#37** $20   **#38** $9   **#39** 50 miles

#40 (1) 4 (2) 5 (3) -3 (4) 4 (5) 12 (6) 10 (7) -5 (8) 0

#41 (1) $\dfrac{1}{6}$ (2) -4 and 6 (3) $\dfrac{7}{2}$

Chapter 4. Inequalities

#1 (1) $a < -3$ (2) $a \geq 2$ (3) $-1 < a \leq 1$ (4) $2a + 3 > \frac{1}{2}a$ (5) $3a - 4 \geq a + 2$ (6) $a \leq 0$

#2 (1) $x > 11$ (2) $x > -4$ (3) $x > \frac{1}{2}$ (4) $x > 2$ (5) $x > \frac{7}{2}$ (6) $x < \frac{5}{2}$ (7) $x \geq -6$

(8) $x \geq -27$ (9) $x < -2$ (10) $x > 4$ (11) No solution   (12) $x \leq \frac{3}{2}$ (13) $x \geq -\frac{7}{5}$

(14) $x < -6$ (15) $x > -\frac{7}{2}$ (16) No solution   (17) $x \geq \frac{5}{23}$ (18) $x > \frac{2}{3}$ (19) $x < \frac{3}{a}$

(20) $x \leq -\frac{3}{a}$ (21) $x < -\frac{1}{2}$ (22) All real numbers

#3 See Solutions Manual.

#4 (1) $x < -\frac{7}{2}$ (2) $x \geq -3$

#5 (1) $-1 \leq 2x + 1 \leq 3$ (2) $-5 \leq -3x - 2 \leq 1$ (3) $-3\frac{1}{4} \leq \frac{1}{4}x - 3 \leq -2\frac{3}{4}$

#6 (1) $-2 < y < \frac{2}{3}$ (2) $2 < y < \frac{10}{3}$ (3) $-4 \leq x \leq -1$

#7 10

#8 22, 23, 24

#9 (1) 9 integers   (2) 2 integers   (3) There are no positive integers.

#10 $8 \leq k < \frac{23}{2}$

#11 (1) 6   (2) -6 (3) $\frac{3}{4}$ (4) $-\frac{4}{3}$ (5) -2 (6) $-\frac{1}{4}$ (7) -3 (8) $-\frac{8}{3}$

#12 $x < -\frac{5}{3}$

#13 See Solutions Manual.   **#14** 29 peaches   **#15** At least 25 ounces of water

#16 $3\frac{3}{7}$ miles   **#17** less than $\frac{3}{4}$ mile   **#18** More than 70 ounces

#19 At least 99 points   **#20** After 21 years   **#21** 2

#22 2   **#23** More than $25   **#24** At least 2 boys

Chapter 5. Functions

#1 $\{-3, -1, 1, 3\}$ **#2** $\{-4, 0, 2, 4\}$

#3 The set of all real numbers that are greater than or equal to -3

#4 $\{-4, 0, 4\}$ **#5** No   **#6** -12 **#7** $\frac{1}{4}$ **#8** -4 **#9** $-\frac{1}{3}$

#10 (1) Not a function   (2) Not a function   (3) A function   (4) A function   (5) A function

    (6) Not a function

#11 1   **#12** $\frac{1}{8}$ **#13** See Solutions Manual.   **#14** See Solutions Manual.

#15 3

#16 (1) $AB = 4$ (2) $CD = 3$ (3) $PQ = 8$ (4) $ST = 6$

#17 (1) III   (2) III   (3) IV   (4) IV   (5) I   (6) II   (7) II   (8) I

#18 24   **#19** 4   **#20** (3), (5), and (7)   **#21** $y = -\frac{4}{3}x$

#22 (1) $y = \frac{5}{3}x$ (2) $y = -\frac{3}{2}x$ (3) $y = -\frac{12}{x}$ (4) $y = \frac{10}{x}$

#23 (1) $y = \frac{4}{x}$ (2) $y = 2x + 3$ (3) $y = \frac{12}{x}$ (4) $y = -\frac{1}{2}x$

#24 $\frac{3}{7}$ **#25** -3 **#26** -12 **#27** -9 **#28** 21   **#29** -8

#30 $y = -\frac{5}{3}x$ or $y = \frac{6}{x}$ **#31** $A(a, 3a) = A(6, 18)$ **#32** 18 square units

#33 $y = \frac{5}{x}$ **#34** (3)   **#35** 25 workers   **#36** $y = \frac{200}{x}$

#37 $\frac{1}{6}$ hour (10 minutes)   **#38** $\frac{5}{12}$ hour (25 minutes)   **#39** 20 miles

#40 $26\frac{2}{3}$ miles per hour   **#41** $y = \frac{300}{x}$ **#42** $20

#43 $3\frac{3}{5}$ hours (3 hours 36 minutes)

Chapter 6. Fractions and Other Algebraic Expressions

#1 (1) hundreds (2) ones (3) tenths (4) ten thousandths (5) thousandths

(6) hundredths (7) tens

#2 (1) 45 , 44.5 , 44.54 , 44.536 (2) 32 , 32.5 , 32.50 , 32.500 (3) 2 , 2.1 , 2.05 , 2.053

(4) 1 , 1.2 , 1.22 , 1.221 (5) 20, 20.0 , 20.00 , 20.000

#3 (1) $-0.24 < -0.024 < 0.05 < 0.418 < 0.48 < 0.5$

(2) $-0.3 < -0.13 < -0.03 < 0.013 < 0.13 < 0.31$

(3) $-0.24 < -0.21 < 2.04 < 2.39 < 2.4 < 2.41$

(4) $0.05 < 0.409 < 0.41 < 0.419 < 0.49 < 0.5$

(5) $-0.61 < -0.6 < -0.59 < -0.509 < -0.061 < -0.06$

#4 (1) 8.14 (2) 6.35 (3) 6.14 (4) 40.138 (5) 6.19 (6) -3.54 (7) 0.55 (8) -5.74

(9) -31.062 (10) 5.21 (11) -14.3 (12) 2.5 (13) -6.1 (14) 3.238 (15) -2.1

(16) 31.09 (17) 0.01 (18) 0 (19) -99.89 (20) -6.22

#5 (1) 0.1 (2) 0.1 (3) 8.5 (4) 0.0099 (5) 40.8 (6) 0.01 (7) -2.08 (8) 13.11

(9) 0.001 (10) 125 (11) 0.4 (12) 300 (13) 0.67 (14) 1.71 (15) 104 (16) 210

(17) -240 (18) -32 (19) 1.25 (20) 100

#6 (1) 1.07 (2) 4.92 (3) -0.02 (4) -0.15 (5) -2.00

#7 (1) $\frac{1}{2}$ (2) $\frac{69}{20}$ (3) $\frac{4}{125}$ (4) $\frac{41}{20}$ (5) $\frac{523}{50}$ (6) $\frac{201}{40}$ (7) $\frac{114}{25}$ (8) $\frac{131}{20}$ (9) $\frac{876}{125}$ (10) $\frac{229}{25}$

#8 (1) $\frac{2}{9}$ (2) $\frac{2}{99}$ (3) $\frac{31}{90}$ (4) $\frac{281}{495}$ (5) $\frac{73}{150}$ (6) $\frac{7}{3}$ (7) $\frac{154}{45}$ (8) $\frac{4241}{495}$ (9) $\frac{1037}{300}$ (10) $\frac{1706}{333}$

#9 (1) 0.375 (2) 0.16 (3) 0.3 (4) 0.075

(5) The fraction cannot be written as a terminating decimal.

(6) The fraction cannot be written as a terminating decimal.

(7) 0.48 (8) 0.035 (9) 0.45 (10) 0.004

#10 (1) $\frac{1}{4}$ (2) $\frac{3}{1}$ (3) $\frac{5}{8}$ (4) $\frac{16}{7}$ (5) $\frac{5}{2}$ (6) $\frac{35}{52}$ (7) $\frac{9}{20}$ (8) $\frac{120}{1}$ (9) $\frac{2}{5}$ (10) $\frac{1}{20}$

#11 (1) 75 ¢   (2) 20 nickels   (3) 90 students   (4) $ 2.50   (5) 63 cookies

#12 (1) 10   (2) 4.8   (3) 1   (4) 56   (5) 6   (6) 0.25   (7) 3   (8) 0.2   (9) 14   (10) 0.05

#13 (1) $\frac{3}{10}$ (2) $\frac{6}{5}$ (3) $\frac{1}{200}$ (4) $\frac{19}{75}$ (5) $\frac{7}{200}$ (6) $\frac{3}{4}$

#14 (1) 91   (2) 62.5   (3) 40   (4) 80%   (5) 5.6   (6) 2.5%   (7) 50%

#15 (1) 75% decrease   (2) 300% increase

#16 (1) 25%   (2) 30%   (3) 4%   (4) 6%   (5) 240%   (6) 1%   (7) 625%   (8) 100%

(9) 1.5%   (10) 60%

#17 (1) 0.25   (2) 0.025   (3) 0.0025   (4) 2.5   (5) 0.225   (6) 0.01   (7) 0.1   (8) 1

(9) 0.5   (10) 0.005

#18 (1) 10%   (2) 1250%   (3) 84%   (4) 4520%   (5) 2%   (6) 0.1%   (7) 47.8%   (8) 100%

(9) 1%   (10) 50%

#19 (1) $100   (2) $37.5   (3) $50.625

#20 (1) $2500   (2) $686   (3) $333

#21 (1) $1098.5   (2) $512.63   (3) $358.22   (4) $324.86   (5) $1016.97

#22 (1) 1.25×10^2 (2) 1.25×10^5 (3) 2.5×10^7 (4) -2.4×10^3 (5) 1.25×10^{-3}

(6) 1.25×10^{-6} (7) 2.5×10^4 (8) -2.4×10^{-3}

#23 (1) 3.454×10^3 (2) 4.1×10^{-4} (3) 2.048×10^4 (4) 3.5×10^8

Chapter 7. Monomials and Polynomials

#1 (1) a^9 (2) $x^7 \cdot y^3 \cdot z$ (3) $2^6 x^2 y^4 z^6$ (4) x^{18} (5) x^{12} (6) $-a^{10} \cdot b^{15}$ (7) $24 x^7 y^5 z^9$

(8) $\dfrac{x^2}{y^4}$ (9) a^1 (10) $\dfrac{1}{4} a^{10}$ (11) $a^5 b^7$ (12) $\dfrac{27}{8}$ (13) $\dfrac{1}{a^2 b^4}$ (14) $\dfrac{1}{x^3}$ (15) $\dfrac{a^7}{b^5}$ (16) $\dfrac{8}{25}$

(17) 9^a (18) 27 (19) $\dfrac{1}{4^8}$ (20) $6x^5 y^3$ (21) $-\dfrac{3}{2} x^7 y^{10}$ (22) $\dfrac{9}{8}$ (23) 1

#2 (4), (6), and (8)

#3 (1) $a = 5, b = 15$ (2) $a = 6$ (3) $a = 3$ (4) $b = 5$ (5) $a = 7$ (6) $a = 12, \ b = 4$

(7) $a = 1$ (8) $a = 3, b = 6$ (9) $a = -2$ (10) $a = 3$ (11) $a = 5$ (12) $a = 1$

(13) $a = 3, b = 5$ (14) $a = 2$ (15) $a = 7, b = 11$

#4 $\dfrac{3}{2} ab$

#5 $\dfrac{2}{3}$

#6 -1

#7 $4^{10}, 8^7, \left(\dfrac{1}{2}\right)^{-30}, 2^{32}$

#8 4

#9 96

#10 $5a^3 b^2$

#11 (1) 6 digits   (2) 5 digits   (3) 12 digits

#12 (1) $a + 4b$ (2) $3a^2 + 4a$ (3) $-a^2 + 4a - 2$ (4) $2x + 4y - 3$ (5) $-\dfrac{13}{6} x - \dfrac{11}{3} y$

(6) $-\dfrac{1}{6} x + \dfrac{11}{6} y - \dfrac{7}{6}$ (7) $4b + 5$

#13 3

#14 (1) 5 (2) $\dfrac{17}{6}$ (3) 0

#15 $-7a^2 + 6a - 3$

#16 $\dfrac{9}{8}$

#17 (1) $6a - 2$ (2) $3x - 4$

#18 (1) $-6x^2 - 8xy + 4x$ (2) $x^3 - y^3$ (3) $-\frac{3}{2}a^2 - ab + 2a^2b$ (4) $-a + b$ (5) $-\frac{9}{2}a + 3b$

(6) $3a^2 - 4a + 1$ (7) $-\frac{1}{2}x^3y^3 + \frac{5}{2}x^3y^2$ (8) $-x^2y^3 + 3x^3y^3 - \frac{8}{9}x + \frac{2}{3}y$

(9) $\frac{2x}{y} - 1 - 2y + \frac{3y}{x}$ (10) $2xy + 3y$ (11) $\frac{1}{4}ab^2 - \frac{7}{3}ab$ (12) $-\frac{1}{2}x^2 + \frac{5}{6}x + 2$

#19 $\frac{5}{4}$

#20 (1) $\frac{2}{3}$ (2) $-3\frac{1}{2}$

#21 (1) $\frac{1}{4}a + \frac{1}{2}b - \frac{1}{6}$ (2) $\frac{3}{4}a^2b - \frac{1}{2}ab^2$

#22 $4ab - b^2$

#23 (1) $2x^2 + x - 15$ (2) $6x^3 - 5x^2 - 3x + 2$ (3) $x^2 - \frac{1}{6}x - \frac{1}{6}$ (4) $9 - 4a^2$ (5) $9a^2 - 4b^2$

(6) $4x^2 + 12x + 9$ (7) $4x^2 - 9$ (8) $a^6 - b^6$ (9) $16x^2 + 4x + \frac{1}{4}$

(10) $x^2 + 2xy + y^2 - 4x - 4y + 4$ (11) 9996 (12) 8556 (13) 9801

(14) $x^2 + 4xy + 4y^2 - 9z^2$ (15) $4x^2 - y^2 + 6y - 9$ (16) 8 (17) $8ab$

(18) $a^4 + 2a^3 - 13a^2 - 14a + 24$

#24 (1) $a^2 + \frac{2}{3}ab - \frac{1}{3}b^2$ (2) $4(a^2 + 2ab - 2b^2)$ (3) $12x^2 + 5x - 3$

#25 (1) -8 (2) $2\frac{1}{3}$ (3) -8 (4) -10

#26 (1) 3 (2) 7 (3) -4 (4) 5 (5) 3 (6) -1 (7) 17 (8) $\frac{25}{144}$ (9) 0 (10) $-\frac{1}{2}$

(11) 19 (12) 17 (13) $\frac{17}{4}$ (14) 1 (15) 3 (16) -2 (17) -1 (18) -6

#27 (1) $x = \frac{3}{2}y - 3$ (2) $y = -\frac{1}{2}x + \frac{3}{2}$ (3) $b = 3a + \frac{1}{2}$ (4) $F = \frac{9}{5}C + 32$ (5) $a = -4b$

(6) $a = \frac{bc}{b-c}$ (7) $a = \frac{1}{2}b$ (8) $b = \frac{c}{1-a}$

#28 (1) 9 (2) 15 (3) 6 (4) $\frac{3}{40}$

#29 (1) 0 (2) 0 (3) -1

Chapter 8. Systems of Equations

#1 (1) $(3,1)$ (2) $\left(\frac{5}{4},\frac{1}{2}\right)$ (3) $(9,-11)$ (4) $(4,-1)$ (5) $(3,9)$ (6) $(-3,-6)$

#2 (1) 4 (2) 2 (3) 3 (4) -5

#3 8   **#4** $(-6,5)$ **#5** -1 **#6** -7

#7 (1) $(5,2)$ (2) $(-1,-4)$ (3) $(2,3)$ (4) $\left(\frac{7}{20},-\frac{7}{4}\right)$ (5) $\left(0,-\frac{1}{2}\right)$ (6) No solution

   (7) $(2,2)$ (8) Unlimited number of solutions   (9) $\left(-4,\frac{4}{3}\right)$ (10) $\left(\frac{1}{11},\frac{2}{11}\right)$

   (11) $\left(-3,-\frac{2}{3},0\right)$ (12) $\left(\frac{5}{3},\frac{3}{2}\right)$

#8 -2 **#9** $-\frac{3}{2}$ **#10** $-\frac{3}{2}$ **#11** (1) 6 (2) 5 (3) 3

#12 -9 **#13** 3   **#14** $\frac{5}{11}$ **#15** 20 square inches

#16 8 children   **#17** 3 apples and 2 peaches

#18 64 candies   **#19** 5 quarters and 13 dimes

#20 48   **#21** $x = 27,\ y = 6$ **#22** 6 ounces of water

#23 5 liters of alcohol   **#24** 2.4 miles   **#25** 7 miles long

#26 12 hours   **#27** 42 years old   **#28** 32 minutes

#29 50 miles   **#30** 1.2 miles per hour

#31 The current number of boys in the club is 18 and the current number of girls in the club is 23 .

#32 16 years old

#33 (1) $(3,1)$ (2) No solution   (3) Unlimited number of solutions

Chapter 9. Systems of Inequalities

#1 (1) $-2 \le x < 4$ (2) $x > 0$ (3) $x < -1$ (4) No solution   (5) No solution   (6) $x = -3$

#2 (1) $x \ge -1$ (2) No solution   (3) $-3 < x < 2$ (4) $x < -1$ (5) $2 \le x < 4$ (6) $\frac{4}{5} \le x < 8$

(7) $x > -1$ (8) $\frac{6}{5} < x < \frac{11}{3}$

#3 (1) $-7 \le y < 4\frac{1}{3}$ (2) $y \le -1$ (3) $y \ge 13$ (4) $-\frac{13}{3} < y < \frac{3}{5}$

#4 (1) 5   (2) $-\frac{1}{5}$ (3) 1   (4) 19   (5) $\frac{5}{3}$ (6) -2

#5 (1) $k > 1$ (2) $k \ge -5$ (3) $5 < k \le 6$ (4) $k \le 0$

#6 (1) 7   (2) 0

#7 -14

#8 (1) $-4 < x \le \frac{5}{2}$ (2) -1 (3) -5

#9 22

#10 $x > 6$

#11 $\frac{5}{4} < x < \frac{8}{3}$

#12 8 miles at most

#13 4

#14 (1) $-1 < x < 7$ (2) $x > 3$ (3) $x < -\frac{9}{2}$ or $x > \frac{11}{2}$ (4) (5) (6) (7) (8) (9) (10)

#15 See Solutions Manual

Chapter 10. Linear Functions

#1 (1) o (2) o (3) × (4) × (5) × (6) o (7) × (8) o (9) × (10) ×

#2 (1) 3 (2) 6 (3) −2 (4) 17

#3 (1) −1 (2) −2 (3) $-\frac{5}{2}$ (4) $\frac{5}{2}$ (5) −3   **#4** (1) 10 (2) −2 (3) 2

#5 (1) The x-intercept is 3 and the y-intercept is 3.

(2) The x-intercept is $-\frac{3}{2}$ and the y-intercept is −6.

(3) The x-intercept is −12 and the y-intercept is 6.

#6 (1) 3 (2) 24 (3) 9 (4) 12   **#7** (1) −2 (2) 1 (3) 2 (4) −1 (5) −2

#8 (1) $m = -\frac{2}{3}$ and $b = \frac{4}{3}$ (2) $m = \frac{4}{3}$ and $b = -\frac{5}{3}$ (3) $m = -\frac{1}{3}$ and $b = -\frac{2}{3}$

(4) $m = 3$ and $b = -5$ (5) $m = 2$ and $b = 0$ (6) $m = 0$ and $b = 2$

#9 (1) $2x - y - 3 = 0$ (2) $y - 5 = 0$ (3) $y = -\frac{2}{3}x + \frac{10}{3}$ (4) $2x + y + 6 = 0$

(5) $3x - y - 1 = 0$ (6) $2x + y - 2 = 0$ (7) $x - 2 = 0$ (8) $2x + y + 1 = 0$

(9) $4x - 7y + 20 = 0$ (10) $3x + 5y - 14 = 0$ (11) $y = 8x + 13$ (12) $x - y + 3 = 0$

(13) $8x - 3y - 12 = 0$ (14) $x + 1 = 0$ (15) $y + 4 = 0$

#10 (1) $2x - y - 1 = 0$ (2) $3x + y - 9 = 0$ (3) $x - 2 = 0$ (4) $y - 3 = 0$

(5) $3x + 4y - 18 = 0$ (6) $3x + 2y - 12 = 0$ (7) $3x - y - 3 = 0$ (8) $y - 3 = 0$

(9) $x - 2 = 0$

#11 (1) −1 (2) $\frac{3}{4}$ (3) −3 (4) $\frac{1}{2}$ (5) 4 (6) −3 (7) $\frac{2}{5}$ (8) 2 (9) 9

#12 (1) $\frac{4}{3}$ (2) −1 (3) $\frac{4}{3}$ **#13** (1) $\frac{7}{3}$ (2) 2 (3) 6 (4) $-\frac{9}{4}$

#14 (1) $-\frac{3}{2}$ (2) 1 (3) $-\frac{1}{2}$ (4) $\frac{21}{16}$

#15 (1) $5x + 7 = 0$ (2) $3x - 5 = 0$ (3) $3x + 2y + 8 = 0$ **#16** $\frac{29}{3}$ **#17** 24

#18 $y = 2x + 10$ **#19** $y = -\frac{1}{2}x + 15$ **#20** $y = -\frac{5}{4}x + 4$ $\left(0 \le x \le 3\frac{1}{5}\right)$

#21 See Solutions Manual.

Chapter 11. The Real Number System

#1 (1) 2^{12} (2) 3^{27} (3) 3^{20} (4) 2^{18} (5) 4^{35} (6) 2^{13} (7) 2^{14} (8) 2^{12} (9) $3^5 \cdot 2^{10}$

(10) $2^{10} \cdot 3^3$ (11) 3^3 (12) $2^{-3} \cdot 5^1$ (13) -2^6 (14) 39 (15) -1

#2 (1) 6 and -6 (2) 3 and -3 (3) 5 and -5 (4) 7 and -7 (5) 11 and -11 (6) $\frac{3}{4}$ and $-\frac{3}{4}$

(7) 0.2 and -0.2 (8) 15 and -15 (9) 0.4 and -0.4 (10) $\frac{6}{7}$ and $-\frac{6}{7}$

#3 (1) 6 (2) 9 (3) 10 (4) 1 (5) 0.7 (6) $\frac{1}{2}$ (7) 0 (8) 13 (9) 0.1 (10) $\frac{1}{4}$

#4 (1) $2 > \sqrt{3}$ (2) $\sqrt{2} > 1.3$ (3) $\sqrt{3} > \frac{3}{4}$ (4) $\sqrt{5} < 2.5$ (5) $-\sqrt{\frac{4}{3}} > -\sqrt{\frac{3}{2}}$ (6) $\sqrt{0.5} > \sqrt{0.05}$

(7) $-\sqrt{2} < \sqrt{5}$ (8) $-\sqrt{0.3} > -\sqrt{\frac{3}{4}}$ (9) $1.4 < \sqrt{2} < 1.5$ (10) $-1.6 > -\sqrt{3} > -1.8$

#5 (1) 6 (2) $2\sqrt{6}$ (3) 9 (4) $2\sqrt{15}$ (5) $90\sqrt{10}$ (6) $2\sqrt{70}$ (7) $6\sqrt{6}$ (8) $12\sqrt{5}$ (9) $\frac{\sqrt{42}}{14}$

(10) $\frac{\sqrt{3}}{6}$ (11) $\sqrt{5}$ (12) $\frac{3}{2}$ (13) $\frac{5}{3}\sqrt{15}$ (14) $3\sqrt{2}$ (15) $5\sqrt{2}$

#6 (1) $x = 4$ (2) $x = 0$ and $x = \frac{2}{3}$ (3) There is no solution.   (4) $x = -2$

#7 (1) $6\sqrt{3}$ (2) $5\sqrt{3}$ (3) $5\sqrt{5}$ (4) $-\sqrt{6}$ (5) $-2\sqrt{2}$ (6) $2\sqrt{3}$ (7) $2\sqrt{5} + 6$ (8) $4 + 3\sqrt{6}$

(9) $2\sqrt{15} - 3\sqrt{5} + 4\sqrt{2}$ (10) $4\sqrt{3} + 3\sqrt{6}$ (11) $\sqrt{30} - 3\sqrt{6} - \sqrt{15} + 3\sqrt{3}$ (12) -6

(13) 18 (14) $7 + 2\sqrt{10}$ (15) $58\sqrt{5} - 12\sqrt{30}$ (16) 22 (17) $2 - 2\sqrt{3}$

(18) $5 + 2\sqrt{6} - \sqrt{10} - \sqrt{15}$ (19) $2^{\frac{11}{6}}$ (20) $2^{\frac{2}{3}} \cdot 3^{\frac{1}{2}} \cdot 5^{\frac{5}{6}}$ (21) $2^{-\frac{1}{2}}$ (22) $\frac{3\sqrt{6}}{5}$ (23) $\frac{4}{3}$ (24) $\sqrt{3}$

#8 (1) $\frac{3\sqrt{2}}{2}$ (2) $\frac{5\sqrt{3}}{3}$ (3) $\frac{\sqrt{6}}{2}$ (4) $\frac{2\sqrt{3}}{3}$ (5) $\frac{\sqrt{5}}{3}$ (6) $-\frac{7}{2}\sqrt{3}$ (7) $\frac{2\sqrt{6} - \sqrt{10}}{6}$ (8) $\sqrt{15} - 1$

(9) $\frac{5\sqrt{6}}{6} - \sqrt{5}$ (10) $5\sqrt{6}$

#9 (1) $a + b = 2\sqrt{3}$, $a - b = 2\sqrt{5}$, $ab = -2$, $a^2 + b^2 = 16$

(2) $a + b = -\sqrt{3}$, $a - b = \sqrt{5}$, $ab = -\frac{1}{2}$, $a^2 + b^2 = 4$

(3) $a + b = -8$, $a - b = 2\sqrt{15}$, $ab = 1$, $a^2 + b^2 = 62$

#10 (1) 5 (2) -5 (3) 3 (4) $5^{\frac{2}{3}}$ (5) -5 (6) -10 (7) $-10+5^{\frac{3}{2}}$ (8) $2^{\frac{1}{3}}$ (9) $\dfrac{2^{\frac{1}{6}}}{5^{\frac{1}{6}}}$ (10) 4

#11 (1) -3 (2) 2

#12 (1) $a+\dfrac{1}{a}=\dfrac{22}{7}$, $a-\dfrac{1}{a}=\dfrac{12\sqrt{2}}{7}$ (2) $41-20\sqrt{3}$ (3) $19-6\sqrt{10}$

#13 (1) -33 (2) $1+2\sqrt{2}$ (3) a (4) $4a$ (5) $2a-1$ (6) $-4a+13$ (7) $-\sqrt{2}$ (8) -1

(9) $2b-2c$ (10) $a-b$

#14 (1) $a+b=1+2\sqrt{7}$ and $a-b=11-2\sqrt{7}$

(2) $ab=12-3\sqrt{12}$ and $\dfrac{a}{b}=3+\dfrac{3\sqrt{3}}{2}$

#15 (1) false   (2) false   (3) true   (4) true   (5) false   (6) false   (7) true   (8) false   (9) false

(10) false

#16 (1) $ab=2\sqrt{2}$, $m=(2,3)$ (2) $ab=\sqrt{89}$, $m=\left(\dfrac{1}{2},-1\right)$ (3) $ab=\sqrt{74}$, $m=\left(\dfrac{1}{2},\dfrac{1}{2}\right)$

(4) $ab=\sqrt{22}$, $m=\left(\dfrac{3-\sqrt{2}}{2},\dfrac{3+\sqrt{2}}{2}\right)$ (5) $ab=\sqrt{\dfrac{73}{4}}$, $m=\left(-2,-\dfrac{1}{4}\right)$

#17 (1) $\dfrac{5}{3}$ (2) $\dfrac{5}{3}$ (3) $\dfrac{13}{3}$ (4) $-\dfrac{5}{3}$ (5) $\dfrac{17}{3}$ (6) $-\dfrac{19}{3}$ (7) $-\dfrac{19}{3}$ (8) $\dfrac{7}{3}$ (9) $-\dfrac{7}{3}$ (10) $\dfrac{17}{3}$

#18 (1) $4,\ 0,\ 5,-1,\ 6,-2$ (2) $15,-13,17,-15$

#19 $a+2b$

#20 (1) $\dfrac{3\sqrt{2}+2}{7}$ (2) 1

Chapter 12. Factorization

#1 (1) $1, a+1, b+1, (a+1)(b+1)$

(2) $1, a, b, (x+y), ab, a(x+y), b(x+y), ab(x+y)$

(3) $1, x, 2x+y, x(2x+y)$

#2 (1) $a(a-b+1)$ (2) $ab(a-b)$ (3) $2(2a-5)$ (4) $a(2a+5)(a-1)$

(5) $3a^2b^2(3a^2-a+4)$ (6) $2(x+y)(a-2b)$ (7) $-6(x-2y)(2+3a)$

(8) $4(a-b)(a-b)$ (9) $a^n(1+a^2)$ (10) $3x^n(x^n+4x^{2n}+3)$

#3 (1) $(x-1)^2$ (2) $(3x+1)^2$ (3) $(2x-1)^2$ (4) $(x+5)^2$ (5) $\left(x+\frac{1}{2}\right)^2$ (6) $(x^3+3)^2$

(7) $(x^n-y^n)^2$ (8) $(x+1)(x-1)$ (9) $(x+2y)(x-2y)$ (10) $\left(3x+\frac{1}{2}y\right)\left(3x-\frac{1}{2}y\right)$

(11) $(x+a+6)(x+a-6)$ (12) $(1+x+y)(1-x-y)$ (13) $(x^2+1)(x+1)(x-1)$

(14) $(x^4+y^4)(x^2+y^2)(x+y)(x-y)$ (15) $(3y+4x)(3y-4x)$ (16) $\left(\frac{1}{3}x+\frac{3}{4}y\right)\left(\frac{1}{3}x-\frac{3}{4}y\right)$

(17) $xy(2x+y)(2x-y)$ (18) $(x+1)(x+3)$ (19) $(x+1)(x-5)$ (20) $(x-1)(x-2)$

(21) $(x-4)(x+2)$ (22) $-4xy$ (23) $(3x-2)(x-4)$ (24) $\left(x-\frac{1}{2}\right)\left(x-\frac{1}{3}\right)$

(25) $(x-2)(3x+2)$ (26) $(x+2)(2x-1)$ (27) $(2x+3)(2x-4)$ (28) $2(x-3)(2x+1)$

(29) $(x-3y)(2x+3y)$ (30) $xy(x+4y)(x-4y)$ (31) $\frac{1}{3}\left(x-\frac{3}{x}\right)^2$ (32) $(2x-3y)^2$

(33) $\frac{1}{3}(x+2)(x-3)$ (34) $(x-y)(a^2+b^2)(a+b)(a-b)$ (35) $-3(a-2)(a+1)$

(36) $a(x-2)(3x+1)$ (37) $(a-6)^2$ (38) $(a^4+1)(a^2+1)(a+1)(a-1)$

(39) $(x+3)(x-2)$ (40) $3(x+1)(x-1)$

#4 (1) $\frac{25}{4}$ (2) 4 (3) $\frac{9}{2}$ (4) $\frac{4}{25}$ (5) $\frac{4}{5}$ or $-\frac{4}{5}$ (6) -20 or 20 (7) 8 or -8

(8) 8 or -4 (9) $12y-5$ or $-12y-5$ (10) $\frac{1}{16}x^2$ (11) 31 or -29 (12) $\frac{1}{4}$

(13) $-12\frac{1}{4}$ (14) $6\frac{1}{4}$ (15) 9

#5 (1) -3 (2) -2 (3) -2 (4) -2 (5) $\frac{3}{2}$

#6 (1) $3\frac{1}{3}$ (2) -4 (3) -6 (4) -1

#7 $10x + 14$

#8 $4x + 3$

#9 $5x + 2$

#10 The perimeter of A is $12x + 16$ and the perimeter of C is $18x + 42$.

#11 250

#12 (1) $-ab(a-2)(a-1)$ (2) $-3(a-2)(a+1)$ (3) $2b(a+2)(4a-5)$

(4) $a(a^2+4)(a+2)(a-2)$ (5) $(x-y)(a+b)(a-b)$ (6) $a(a-b)(2a-b)$

(7) $(a-b)^2(1+2a-2b)$ (8) $(x-y)(ax-ay+b)$ (9) $x(4x+13)$

(10) $(x-y-1)(x-y+4)$ (11) $(2x+3y)^2$ (12) $(x-3y-4)(2x+3y+1)$

(13) $(a+3)(a-3)(a^2+4)$ (14) $(a^2+2)(a^2-2)(a^4+2)$ (15) $(x-y)(x+2)$

(16) $(a+1)(a-b-1)$ (17) $(a-1)^2(a+1)$ (18) $a(a-3)(a+1)(a-1)$

(19) $(1+a-b)(1-a+b)$ (20) $(x^2-2x-9)(x^2-2x+2)$

(21) $(3x+y+2)(3x-y-2)$ (22) $(a+b-2)(a-b+2)$ (23) $(a+4b-3)(a-4b-3)$

(24) $(a-b)(x+1)(x-1)$ (25) $(x-2)(x-y+3)$ (26) $(2a-1)(b-a-3)$

(27) $(x+y-1)(2x-y)$ (28) $(2a+b-3)(2a-b+1)$ (29) $(2a-b+c)(2a-b-c)$

(30) $(a^2+a+1)(a^2-a+1)$ (31) $(a^2+2a-1)(a^2-2a-1)$

(32) $(a^2+3a-2)(a^2-3a-2)$ (33) $(3x^2+2x+2)(3x^2-2x+2)$

#13 (1) 9800   (2) 1880   (3) -200 (4) 6560   (5) 21   (6) $\frac{101}{200}$ (7) 2^{16} (8) 1   (9) 34

(10) 8700   (11) 900   (12) 1600   (13) 540   (14) 8500

Chapter 13 Quadratic Equations

#1 (1) Yes   (2) Yes   (3) No   (4) No   (5) Yes   (6) No   (7) No   (8) Yes   (9) Yes

(10) No   (11) Yes   (12) Yes   (13) No   (14) No   (15) Yes   (16) Yes   (17) Yes

(18) No   (19) Yes   (20) No

#2 (1) $a \neq 2$ (2) $a \neq -2$ (3) $3a + b \neq 0$ (4) $ab \neq 6$ (5) $2a + b \neq 0$ (6) $a \neq 0$

#3 (1) -9 (2) -5 (3) 6   (4) 0   (5) 1

#4 (1) 2   (2) $3\frac{1}{2}$ (3) 2   (4) -1 (5) -5 (6) $\frac{1}{2}$

#5 (1) $a = 3,\ \beta = -4$ (2) $a = -\frac{3}{2},\ \beta = \frac{1}{2}$ (3) $a = -3,\ \beta = \frac{3}{2}$ (4) $a = -\frac{1}{4},\ \beta = -6$

(5) $a = 2,\ \beta = -\frac{3}{2}$

#6 (1) 0   (2) $4\frac{1}{4}$ (3) 13   (4) $6\frac{4}{9}$ (5) $-\frac{4}{3}$ (6) 22

#7 (1) 5   (2) 7   (3) $-\frac{5}{13}$ (4) $\frac{5}{36}$

#8 (1) ± 2 (2) $\pm \frac{\sqrt{5}}{3}$ (3) $1 \pm \sqrt{5}$ (4) $\frac{-5 \pm \sqrt{3}}{2}$ (5) $2 \pm \frac{1}{2}$

#9 (1) $\frac{3}{2} \pm \frac{\sqrt{21}}{2}$ (2) $x = 1$ or $x = -\frac{7}{2}$ (3) $x = -\frac{3}{2} \pm \frac{\sqrt{29}}{2}$ (4) $x = 1$ or $x = \frac{1}{3}$

#10 (1) 0   (2) $\pm 2\sqrt{5}$ (3) ± 1 (4) $\frac{9}{8}$ (5) $6\frac{1}{8}$ (6) 2   (7) $-1 \pm \frac{4\sqrt{6}}{3}$

#11 (1) $\frac{53}{16}$ (2) $2\frac{1}{3}$ (3) 4

#12 (1) $a < -2$ (2) $a < \frac{3}{4}$ (3) $a = \frac{1}{8}$ (4) $a > \frac{1}{12}$ (5) $a = \pm\sqrt{5}$

#13 (1) $7\frac{1}{2}$ (2) -8 (3) $\frac{7}{9}$

#14 (1) $a = -6,\ b = -3$ (2) $a = -3,\ b = 9$ (3) $a = -2, b = -1$ (4) $a = 8$ and $b = 0$

#15 (1) $1 \pm \sqrt{5}$ (2) $\frac{-5 \pm \sqrt{37}}{6}$ (3) $\frac{1 \pm \sqrt{6}}{5}$ (4) -1 or $\frac{5}{2}$ (5) $3 \pm \sqrt{5}$ (6) $\frac{3 \pm \sqrt{3}}{2}$ (7) $\frac{1 \pm \sqrt{21}}{2}$

(8) $\frac{-7 \pm \sqrt{17}}{2}$ (9) $\frac{8 \pm \sqrt{19}}{5}$ (10) $-3 \pm \frac{\sqrt{10}}{2}$

#16 (1) 0   (2) -2 (3) $\frac{9}{2}$ (4) $\frac{13}{30}$ (5) $\frac{17}{15}$ (6) $\frac{2}{7}$

#17 (1) 2 different solutions   (2) No solution   (3) Only one solution   (4) 2 different solutions

(5) 2 different solutions   (6) No solution

#18 (1) $a > 6$ (2) $a < -3$ or $-3 < a < \frac{3}{2}$ (3) $a = \frac{-4 \pm 2\sqrt{31}}{9}$ (4) $a = 2 + 2\sqrt{3}$

(5) $a = -\frac{1}{2}$ (6) $a < 1$

#19 (1) $-\frac{5}{3}$ (2) $\frac{37}{9}$ (3) $\pm \frac{7}{3}$ (4) $-\frac{35}{9}$ when $\alpha - \beta = \frac{7}{3}$; $\frac{35}{9}$ when $\alpha - \beta = -\frac{7}{3}$ (5) $\frac{5}{2}$

#20 (1) $-3k$ (2) $2k^2 - 4k - 1$ (3) $5k^2 + 8k + 2$ (4) $k^2 + 16k + 4$ (5) $\frac{5k^2 + 8k + 2}{2k^2 - 4k - 1}$

(6) $k = -\frac{1}{2}$ or $k = 1$

#21 (1) $x = -3$ or $x = 2$ (2) No solution   (3) $x = \frac{3 \pm \sqrt{41}}{-4}$ (4) $x = -\frac{4}{3}$ or $x = -2$

(5) No solution

#22 46

#23 $x^2 + 5x + 6 = 0$

#24 (1) $0 < a < \frac{9}{8}$ (2) $0 < a < \frac{1}{3}$ (3) $a < 0$

#25 (1) $x = -3$ (when $k = 6$) or $x = -1$ (when $k = 2$)   (2) $x = -\frac{1}{2}$

(3) $x = -2$ (when $k = 2$) or $x = -\frac{4}{3} \left(\text{when } k = \frac{2}{3} \right)$

#26 8-sided polygon

#27 7   **#28** 20   **#29** 25 and 9   **#30** $x = \frac{5}{2}$ **#31** 32

#32 32 inches   **#33** 3 inches

#34 (1) $t = 2$ seconds (upward) and $t = 10$ seconds (downward)   (2) $t = 12$ seconds

(3) 160 feet   (4) 180 feet

Chapter 14. Rational Expressions (Algebraic Functions)

#1 (1) $2x^3$　　(2) $\frac{3x^2}{14y^5}$, $x \neq 0$　　(3) $-\frac{2}{3}$, $x \neq 4$　　(4) $\frac{x+2}{x-1}$, $x \neq 4$　　(5) $\frac{7}{2x}$

　　(6) $\frac{-x+6}{(x+2)(x-2)}$　　(7) $-\frac{x+5}{(x+3)(x-3)(x-5)}$　　(8) $\frac{1}{(x-2)(x-1)}$, $x \neq -3$, $x \neq -2$

　　(9) $\frac{(x-4)(x-3)}{(x+5)}$, $x \neq -4$　　(10) $\frac{x^2}{4}$　　(11) $\frac{2x}{x-4}$, $x \neq 3$, $x \neq -\frac{1}{2}$

#2 (1) $x + 7 + \frac{8}{x-2}$　　(2) $2x - 6 - \frac{2}{x+3}$　　(3) $3x^2 + 8x + 16 + \frac{33}{x-2}$

#3 (1) $x = \frac{5}{12}$　　(2) $x = 0$　　(3) $x = -3$ or $x = \frac{1}{2}$　　(4) $x = -\frac{8}{3}$

#4 (1) $x < -3$ or $x > 4$　　(2) $-5 \leq x < 2$　　(3) $-2 \leq x < 3$ or $x \geq 5$

　　(4) $-\frac{1}{2} < x < 3$ or $x > 4$

Chapter 15. Quadratic Functions

#1 (1) O   (2) ×   (3) ×   (4) ×   (5) O   (6) O   (7) O   (8) ×   (9) ×   (10) ×

#2 (1) −1   (2) 2   (3) 6   (4) $\frac{1}{4}$ (5) −4

#3 (1) −7   (2) $\frac{15}{8}$ (3) 2

#4 (1) Vertex : $(1, -2)$ and axis of symmetry : $x = 1$

(2) Vertex : $(1, -4)$ and axis of symmetry : $x = 1$

(3) Vertex : $(-1, 3)$ and axis of symmetry : $x = -1$

(4) Vertex : $(0, 1)$ and axis of symmetry : $x = 0$

(5) Vertex : $(2, 3)$ and axis of symmetry : $x = 2$

#5 (1) $y = \frac{1}{2}(x + 2)^2$ (2) $y = \frac{1}{2}x^2 + 2$ (3) $y = \frac{1}{2}(x - 1)^2 - 1$ (4) $y = \frac{1}{2}(x + 3)^2 - 4$

(5) $y = \frac{1}{2}(x - m)^2 + n$

#6 (1) $-\frac{1}{2}$ (2) $\frac{3}{4}$ (3) 10   (4) −6   (5) $\frac{1}{3}$

#7 (1) $\frac{5}{4}$ (2) 8

#8 (1) −9   (2) 1   (3) 4

#9 (1) $m > 0,\ n < 0$ (2) $m > 0,\ n = 0$ (3) $m < 0,\ n = 0$ (4) $m = 0,\ n < 0$

(5) $m < 0,\ n > 0$ (6) $m < 0,\ n < 0$ (7) $m > 0,\ n > 0$

#10 (1) $y = 2\left(x - \frac{7}{2}\right)^2 - \frac{1}{2}$ (2) $y = -3\left(x + \frac{5}{3}\right)^2 + \frac{4}{3}$ (3) $y = \frac{1}{2}(x + 2)^2 + 1$

(4) $y = \frac{1}{2}(x - 2)^2 + 1$

#11 $A: y = \frac{1}{2}(x + 5)^2$, $B: y = -\frac{1}{2}(x - 3)^2 - 2$

#12 (1) $\frac{1}{2}$ unit along the x-axis and $\frac{27}{4}$ units along the y-axis

(2) $-\frac{5}{4}$ units along the x-axis and $-\frac{19}{16}$ units along the y-axis

(3) $\frac{1}{2}$ unit along the x-axis and $\frac{15}{4}$ units along the y-axis

(4) 4 units along the y-axis

#13 (1) Vertex : $\left(-\dfrac{3}{4}, -\dfrac{1}{8}\right)$ and axis of symmetry : $x = -\dfrac{3}{4}$

x–intercepts are $x = -\dfrac{1}{2}$ and $= -1$; y–intercept: $y = 1$

(2) Vertex : $(1, 4)$ and axis of symmetry : $x = 1$

x–intercepts are $x = 3$ and $x = -1$; y–intercept: $y = 3$

(3) Vertex : $\left(-\dfrac{1}{2}, \dfrac{3}{4}\right)$ and axis of symmetry : $x = -\dfrac{1}{2}$

x–intercepts are $x = 0$ and $x = -1$; y–intercept: $y = 0$

(4) Vertex : $(4, -2)$ and axis of symmetry : $x = 4$

x–intercepts are $x = 6$ and $x = 2$; y–intercept: $y = 6$

#14 (1) $\dfrac{5}{3}$ (2) -3 (3) -5

#15 (1) $y = (x - 1)^2 + 2$ (2) $y = -2(x + 1)^2 + 6$ (3) $y = -(x + 1)^2$

(4) $y = -\dfrac{7}{8}\left(x - \dfrac{11}{7}\right)^2 - \dfrac{47}{56}$ (5) $y = \dfrac{3}{8}(x - 3)^2 - \dfrac{27}{8}$ (6) $y = \dfrac{1}{3}\left(x - \dfrac{3}{2}\right)^2 - \dfrac{27}{4}$

(7) $y = \dfrac{1}{2}\left(x + \dfrac{5}{2}\right)^2 - \dfrac{1}{8}$

#16 (1) $y = -\dfrac{1}{2}(x - 2)^2$ (2) $y = (x - 2)^2 - 1$ (3) $y = -\dfrac{2}{5}(x + 3)^2 + \dfrac{8}{5}$

(4) $y = \dfrac{1}{3}(x + 1)^2 - \dfrac{16}{3}$ (5) $y = -\dfrac{3}{4}(x - 2)^2 + 3$

#17 (1) $a = -4$ (2) $a = 9$ (3) $a = -8$

#18 (1) $y = 1$ is the minimum value at $x = 2$.   (2) $y = 3$ is the maximum value at $x = 1$.

(3) $y = 12$ is the maximum value at $x = 1$.   (4) $y = 0$ is the maximum value at $x = 2$.

#19 (1) $y = \dfrac{2}{9}(x - 1)^2 + 3$ (2) $y = -3(x + 1)^2 + 4$

#20 (1) 4   (2) -3 (3) $\dfrac{1}{2}$ (4) 1

#21 $\dfrac{25}{4}$ square inches

#22 81   #23 -25 #24 17 feet

Chapter 16. Basic Statistical Graphs

#1 (1) The most: English, the least: Science   (2) Math and Music

(3) 5 students   (4) 20%

#2 See Solutions Manual.

#3 See Solutions Manual.

#4 See Solutions Manual.

#5 (1) Tuesday   (2) Friday and Sunday   (3) Tuesday and Wednesday

Chapter 17. Descriptive Statistics

#1 See Solutions Manual.

#2 See Solutions Manual.

#3 See Solutions Manual.

#4 (1) Mean = 4.286, Median = 4, Mode = 3, Range = 5

(2) Mean = 6.75, Median = 5, Mode = 5, Range = 18

(3) Mean = 26.7, Median = 25, No Mode, Range = 17

(4) Mean = -4.75, Median = -4.5, Mode = -4 and -7, Range = 5

#5 The deviations are $-13, -8, -2, 0, 5, 18$.

The variance is $S^2 = 97.67$

The standard deviation is $s = \sqrt{97.67} \approx 9.89$

#6 (1) $Q_L = 27$ and $Q_U = 39$ (2) $IR = 12$ (3) See Solutions Manual.

#7 See Solutions Manual.

#8 See Solutions Manual.

Chapter 18. The Concept of Sets

#1 (1) × (2) × (3) o (4) × (5) × (6) o (7) × (8) o (9) o (10) ×

#2 (1) True (2) True (3) True (4) False (5) True (6) True (7) True (8) False

#3 (1) Finite (2) Infinite (3) Infinite (4) Finite (5) Finite (6) Infinite

#4 (1) 8 (2) 4 (3) 0 (4) 8

#5 (1) 2 (2) 3

#6 (1) 15 (2) 7

#7 4

#8 8

#9 (1) 4 (2) $A = \{2, 3, 4\}$ or $\{2, 3, 5\}$ or $\{2, 3, 6\}$ (3) 12 (4) 48

#10 (1) 9 (2) 7 (3) 8

#11 (1) $\{2, 6\}$ (2) $\{2, 3, 4, 5\}$ (3) $\{3, 6, 12\}$ (4) $\emptyset$

#12 (1) $\{4, 5\}$ (2) $\{1, 2, 5, 6\}$ (3) $\{1, 2, 4, 5\}$ or $\{1, 3, 4, 5\}$ or $\{2, 3, 4, 5\}$ (4) $\{2, 3, 4\}$

#13 (1) 4 (2) 8 (3) 8 (4) 12 (5) 4 (6) 32

#14 (1) $A - B = \{3, 5\}$ and $B - A = \emptyset$

(2) $A - B = \{2, 4, 6\}$ and $B - A = \{7\}$

(3) $A - B = \{2\}$ and $B - A = \{4, 20\}$

#15 (1) $U - A$ (2) A (3) A (4) A (5) U (6) $A - B$ (7) $U - (A \cup B)$ (8) $\emptyset$ (9) A

(10) $\emptyset$ (11) $B - A$

Chapter 19. Probability

#1 (1) $\{0, 1, 2, 3\}$

(2) {H0, H1, H2, H3, T0, T1, T2, T3}

(3) {HH, HT, TH, TT}

(4) $\{11, 12, \cdots, 16, 21, 22, \cdots 26, \cdots \cdots, 61, 62, \cdots, 66\}$

(5) $\{H1, H2, \cdots, H6, T1, T2, \cdots, T6\}$

#2 (1) 0   (2) $\frac{1}{4}$ (3) 1

#3 (1) $P(\text{red}) = P(\text{black}) = \frac{26}{52}$

(2) $P(\text{ace or picture card}) = \frac{16}{52}$, $P(\text{otherwise}) = \frac{36}{52}$

#4 (1) $P(\text{red}) = \frac{3}{9} = \frac{1}{3}$, $P(\text{blue}) = \frac{4}{9}$, $P(\text{yellow}) = \frac{2}{9}$

(2) $P(\text{even}) = \frac{4}{9}$, $P(\text{odd}) = \frac{5}{9}$

#5 (1) $\frac{1}{8}$ (2) $\frac{1}{4}$ (3) $\frac{1}{8}$ (4) $\frac{1}{8}$ (5) $\frac{1}{4}$

#6 (1) 1   (2) $\frac{1}{2}$ (3) $\frac{1}{18}$ (4) $\frac{5}{36}$ (5) $\frac{11}{12}$

#7 (1) 5,040   (2) 362,880

#8 $\frac{1}{2}$

#9 $\frac{1}{3}$

#10 (1) $\frac{3}{20}$ (2) $\frac{2}{5}$ (3) $\frac{3}{8}$

#11 $P(E \cup F) = \frac{7}{12}$, $P(E \cap F) = \frac{1}{12}$, E and F are independent.

#12 E and F are independent.

Index

A

B

C

O

P

Q

R

S

T

U

V

Z

Made in the USA
Columbia, SC
17 May 2023

16875100R10172